CONTENTS

THE UNIFIED MATH FRAMEWORK: A COMPREHENSIVE GUIDE TO ALGEBRA, GEOMETRY, AND TRIGONOMETRY

By

Tony Yustein © 2024
https://thecode.wiki

CHAPTER 1: THE HISTORICAL DEVELOPMENT OF MATHEMATICS

• •
Mathematics is the universal language of science, art, and nature. It has shaped human civilization by enabling us to understand patterns, structures, and the laws of the universe. To truly appreciate the beauty of algebra, geometry, and trigonometry, we must first explore their origins, key figures, and the profound evolution of mathematical thought.

• •

ORIGINS OF ALGEBRA, GEOMETRY, AND TRIGONOMETRY

Algebra: The Art of Abstraction

Algebra, derived from the Arabic word *al-jabr* (meaning "reunion of broken parts"), originated in ancient Mesopotamia around 2000 BCE. The Babylonians developed the earliest known algebraic methods to solve quadratic equations, using systematic approaches recorded on clay tablets.

- **Babylonian Contributions**:

  - Tables of squares, cubes, and reciprocal numbers were the precursors to algebraic computations.

  - They solved equations such as $ax^2 + bx = c$, albeit without formal notation.

The discipline was revolutionized in the Islamic Golden Age by **Al-Khwarizmi** (circa 780–850 CE), known as the "father of algebra." His seminal work, *Al-Kitab al-Mukhtasar fi Hisab al-Jabr wal-Muqabala* (The Compendious Book on Calculation by Completion and Balancing), formalized algebra as a distinct mathematical field. His methods for solving linear and quadratic equations laid the foundation for modern algebra.

Geometry: The Language of Space

Geometry, one of the oldest branches of mathematics, originated from practical needs such as land measurement, architecture, and astronomy.

- **Egyptian and Mesopotamian Roots**:

  - Ancient Egyptians used geometric principles to construct pyramids and redivide land after annual flooding.

  - Mesopotamians applied geometry in astronomical calculations and city planning.

The field reached its zenith with **Euclid** (circa 300 BCE), often called the "father of geometry." His work, *Elements*, systematically compiled the known geometric knowledge of his time. It established axiomatic geometry, where logical deductions are made from a set of basic postulates. *Elements* remained a cornerstone of mathematical education for over two millennia.

Trigonometry: The Study of Angles and Ratios

Trigonometry's origins are intertwined with the study of astronomy. Early civilizations, such as the Babylonians and Egyptians, used rudimentary trigonometric concepts to measure celestial bodies.

- **Greek Advances**:

  - The Greeks formalized trigonometry as a mathematical discipline.

  - **Hipparchus** (circa 190–120 BCE), often called the "father of trigonometry," created the first known trigonometric table.

- **Indian Contributions**:

  - Indian mathematicians like **Aryabhata** and **Brahmagupta** introduced sine functions and advanced trigonometric calculations.

The field flourished in the Islamic world, where scholars like **Al-Battani** refined trigonometric tables, introducing tangent and cotangent functions. These advancements were critical to the navigation and astronomy of the time.

. . . .

KEY FIGURES IN MATHEMATICS

Euclid: The Father of Geometry

- **Major Contributions**:

  - Formalized the axiomatic approach in *Elements*.

  - Defined foundational concepts such as points, lines, and planes.

- **Impact**:

  - Inspired generations of mathematicians, including Isaac Newton and René Descartes.

  - Provided a framework for logical reasoning in mathematics and science.

Al-Khwarizmi: The Pioneer of Algebra

- **Major Contributions**:

  - Introduced systematic methods for solving equations.

  - Developed the Hindu-Arabic numeral system, a precursor to modern positional notation.

- **Impact**:

  - His name gave rise to the term "algorithm," a cornerstone of computer science.

  - Made mathematics more accessible and practical for diverse fields like commerce and engineering.

Pythagoras: The Philosopher Mathematician

- **Major Contributions**:
  - Established the Pythagorean theorem, linking algebra and geometry ($a^2 + b^2 = c^2$).
  - Founded a mystical school that viewed numbers as the essence of reality.
- **Impact**:
  - His theorem remains a fundamental principle in mathematics.
  - Inspired mathematical exploration into relationships between numbers, music, and the cosmos.

· ·

THE EVOLUTION OF MATHEMATICAL THOUGHT

From Practical to Abstract

Initially, mathematics was driven by practical needs, such as taxation, land division, and trade. Over time, it evolved into an abstract discipline with universal principles.

- The **Babylonians and Egyptians** focused on computation and measurement.

- The **Greeks** introduced rigor and logic, elevating mathematics to a philosophical endeavor.

- The **Indians** and **Islamic scholars** bridged abstract mathematics and practical applications, enriching the field with innovations in algebra and trigonometry.

Role in Science and Philosophy

Mathematics became the foundation of scientific inquiry during the Renaissance. Galileo Galilei famously stated, "Mathematics is the language with which God has written the universe."

- It played a pivotal role in:

  - **Physics**: Describing motion, forces, and celestial mechanics.

  - **Engineering**: Designing bridges, buildings, and machines.

  - **Philosophy**: Influencing logical thought and

epistemology.

Mathematics in the Modern Era

The development of calculus by Newton and Leibniz, the rise of algebraic structures in the 19th century, and the advent of modern geometry and trigonometry have cemented mathematics as the foundation of technology and innovation.

• •

CONCLUSION

The historical development of mathematics is a testament to humanity's quest for understanding and mastery over the natural world. The origins of algebra, geometry, and trigonometry, shaped by luminaries like Euclid, Al-Khwarizmi, and Pythagoras, illustrate the enduring power of mathematical thought. These disciplines not only solved practical problems but also unlocked the secrets of the universe, paving the way for advances in science, technology, and philosophy.

CHAPTER 2: FUNDAMENTALS OF GEOMETRY

• •

Geometry, the branch of mathematics concerned with the properties and relationships of shapes and spaces, is one of the most ancient and essential disciplines in mathematics. Its principles are deeply woven into the fabric of the natural world, architecture, engineering, and art. This chapter introduces the fundamental concepts of geometry, explores its core subfields, and reveals its ubiquitous presence in nature and design.

• •

CORE CONCEPTS
OF GEOMETRY

1. Points

- **Definition**: A point represents a specific location in space but has no size, area, or dimension. It is the most basic unit of geometry.

- **Notation**: Denoted by a single capital letter (e.g., A, B).

- **Applications**: Points are used to define locations, such as the vertices of polygons or coordinates in a plane.

2. Lines

- **Definition**: A line is an infinite set of points extending in both directions without width or thickness.

- **Types of Lines**:

  - **Straight Line**: Infinite and unchanging in direction.

  - **Line Segment**: A portion of a line bounded by two endpoints.

  - **Ray**: A line with one endpoint, extending infinitely in the other direction.

- **Notation**: Represented by two points on the line (e.g., $\overleftrightarrow{AB}$).

- **Applications**: Lines are essential for defining angles, polygons, and intersections.

3. Angles

- **Definition**: An angle is formed when two rays share a

common endpoint (the vertex).

- **Types of Angles**:

  - **Acute Angle**: Less than 90°.

  - **Right Angle**: Exactly 90°.

  - **Obtuse Angle**: Greater than 90° but less than 180°.

  - **Straight Angle**: Exactly 180°.

- **Notation**: Denoted by the vertex and two points on the rays (e.g., $\angle ABC$).

- **Applications**: Angles are central to geometry, appearing in triangles, polygons, and circular arcs.

4. Shapes

- **Definition**: A shape is a two-dimensional geometric figure bounded by lines or curves.

- **Types of Shapes**:

  - **Polygons**: Closed figures with straight sides (e.g., triangles, quadrilaterals, pentagons).

  - **Circles**: A set of points equidistant from a central point.

- **Properties**:

  - **Perimeter**: The total length of a shape's boundary.

  - **Area**: The space enclosed within a shape.

- **Applications**: Shapes form the basis of architectural design, art, and computational modeling.

5. Solids

- **Definition**: Solids are three-dimensional geometric objects.

- **Types of Solids**:

  - **Polyhedra**: Solids with flat faces (e.g., cubes, pyramids).

  - **Curved Solids**: Solids with curved surfaces (e.g., spheres, cylinders).

- **Properties**:

  - **Surface Area**: The total area of all external surfaces.

  - **Volume**: The space occupied by the solid.

- **Applications**: Solids are foundational in 3D modeling, construction, and physics.

• •

INTRODUCTION TO EUCLIDEAN AND NON-EUCLIDEAN GEOMETRIES

Euclidean Geometry

- **Overview**: Named after the ancient Greek mathematician Euclid, it is based on five postulates outlined in his work *Elements*.

- **Key Postulates**:

  1. A straight line can be drawn between any two points.

  2. A finite straight line can be extended infinitely in a straight line.

  3. A circle can be drawn with any center and radius.

  4. All right angles are equal.

  5. The parallel postulate: Through a point not on a line, there is exactly one parallel line.

- **Applications**:

  1. Basis for most classical geometry.

  2. Used in fields such as architecture, engineering, and art.

Non-Euclidean Geometry

- **Overview**: Developed in the 19th century, it modifies or rejects Euclid's parallel postulate.

- **Types**:

  - **Hyperbolic Geometry**: Multiple parallel lines can pass through a point not on a line.

  - **Elliptic Geometry**: No parallel lines exist; all lines eventually intersect.

- **Applications**:

  - Hyperbolic geometry is used in models of the universe and relativity theory.

  - Elliptic geometry is used in navigation and astronomy.

Comparison of Euclidean and Non-Euclidean Geometry

Feature	Euclidean Geometry	Non-Euclidean Geometry
Parallel Lines	Exactly one through a point	Many (hyperbolic) or none (elliptic)
Space Curvature	Flat	Curved (positive or negative)
Real-world Use	Everyday measurements	Advanced physics and astronomy

• •

GEOMETRY IN NATURE AND DESIGN

Geometry in Nature

- **Examples**:

    - **Symmetry**: Found in flowers, snowflakes, and animals.

    - **Fractals**: Self-similar patterns, such as in ferns and coastlines.

    - **Golden Ratio**: Appears in the arrangement of leaves, shells, and galaxies.

- **Applications**:

    - Explaining natural growth patterns.

    - Modeling ecosystems and environmental phenomena.

Geometry in Design

- **Architecture**:

    - Use of geometric principles in designing buildings, domes, and bridges.

    - Examples: The Pyramids of Giza, the Parthenon, and modern skyscrapers.

- **Art**:

    - Symmetry and proportion in works by Leonardo da Vinci and M.C. Escher.

- **Technology**:

    - Computer-aided design (CAD) relies on geometric

principles.

- Applications in 3D printing and animation.

• •

CONCLUSION

Geometry serves as a bridge between the tangible and the abstract, enabling us to describe, analyze, and create structures in two and three dimensions. From the foundational elements of points, lines, and angles to the expansive realms of Euclidean and non-Euclidean geometries, this discipline is both ancient and modern. Its principles pervade the natural world and human design, affirming geometry's timeless significance.

CHAPTER 3:
FUNDAMENTALS
OF ALGEBRA

Algebra, often referred to as the "language of mathematics," provides the foundation for expressing and solving problems in a symbolic form. By using variables, constants, and equations, algebra allows us to generalize mathematical relationships and apply them to diverse fields such as physics, economics, and computer science. This chapter explores the fundamental concepts of algebra, delves into linear and quadratic equations, and illustrates how algebra serves as a universal tool for mathematical communication.

CORE CONCEPTS
OF ALGEBRA

1. Variables

- **Definition**: A variable is a symbol, typically a letter, that represents a number whose value can change or is unknown.

- **Examples**:

    - x, y, z: Commonly used to represent unknowns in equations.

    - t: Often represents time in equations.

- **Applications**:

    - Variables allow for generalization in equations and formulas. For example, $A = l \times w$ represents the area of a rectangle for any length l and width w.

2. Constants

- **Definition**: A constant is a fixed value that does not change.

- **Examples**:

    - Numbers like $5, -3, \frac{1}{2}, \pi,$ and e.

    - In $y = mx + b$, b is a constant that represents the y-intercept of a line.

- **Applications**:

  – Constants are used to define specific properties or quantities in equations.

3. Basic Operations

Algebra relies on basic mathematical operations to manipulate expressions and solve equations.

- **Addition and Subtraction**:

  – Combine or separate terms (e.g., $2x + 3x = 5x$).

- **Multiplication and Division**:

  – Simplify terms (e.g., $3x \times 2 = 6x$, $\dfrac{6x}{3} = 2x$).

- **Exponentiation**:

  – Powers and roots (e.g., x^2, $\sqrt{x}$).

- **Order of Operations**:

  – Follow the rules of **PEMDAS**: Parentheses, Exponents, Multiplication/Division, Addition/Subtraction.

• •

LINEAR AND QUADRATIC EQUATIONS

1. Linear Equations

- **Definition**: An equation that forms a straight line when graphed. It has the general form: $y = mx + b$ where:

  - m is the slope (rate of change).

  - b is the y-intercept (value of y when $x = 0$).

- **Examples**:

  - $2x + 3 = 7$: Solves to $x = 2$.

  - $y = 4x - 5$: Represents a line with slope 4 and y-intercept -5.

- **Graphical Representation**:

  - A straight line that rises, falls, or remains constant, depending on the slope.

2. Quadratic Equations

- **Definition**: An equation that forms a parabola when graphed. It has the general form: $y = ax^2 + bx + c$ where:

  - a, b, and c are constants, and $a \neq 0$.

- **Examples:**

  - $x^2 - 4x + 3 = 0$: Solves to $x = 1$ and $x = 3$.

- **Graphical Representation:**

  - A parabola that opens upward if $a > 0$ or downward if $a < 0$.

- **Key Features:**

  - **Vertex:** The highest or lowest point on the parabola.

  - **Axis of Symmetry:** A vertical line that divides the parabola into two symmetric halves ($x = -\frac{b}{2a}$).

  - **Roots (Solutions):** The x-values where the parabola intersects the x-axis.

• •

GRAPHICAL ANALYSIS

1. Graphing Linear Equations

- **Steps**:

  1. Identify the slope (m) and y-intercept (b).
  2. Plot the y-intercept on the graph.
  3. Use the slope to determine another point on the line.
  4. Connect the points to form the line.

- **Example**:

  1. For $y = 2x + 1$, the slope is 2, and the y-intercept is 1. Plot $(0,1)$, then move up 2 and right 1 to find another point, such as $(1,3)$.

2. Graphing Quadratic Equations

- **Steps**:

  1. Identify the coefficients $a, b,$ and c.
  2. Find the vertex using $x = -\dfrac{b}{2a}$.
  3. Determine additional points by substituting values of x into the equation.
  4. Sketch the parabola.

- **Example**:

1. For $y = x^2 - 4x + 3$, the vertex is $(2, -1)$. Plot additional points, such as $(1,0)$ and $(3,0)$, to complete the parabola.

· ·

ALGEBRA AS THE LANGUAGE OF MATHEMATICS

1. Generalization of Patterns

- Algebra allows the representation of patterns and relationships in a concise, universal form.

- Example: The sum of an arithmetic sequence can be expressed as: $S_n = \frac{n}{2}(a + l)$ where n is the number of terms, a is the first term, and l is the last term.

2. Problem Solving

- Algebraic equations enable the formulation and solution of real-world problems.

- Example: Calculating interest using the formula: $A = P(1 + rt)$ where A is the total amount, P is the principal, r is the interest rate, and t is time.

3. Mathematical Modeling

- Algebra is essential for creating models that describe physical, economic, and social systems.

- Example: Modeling population growth with exponential functions: $P(t) = P_0 e^{rt}$

4. Communication Across Disciplines

- Algebra provides a common framework for collaboration between fields such as engineering, physics, computer science, and finance.

- Example: Algorithms, which are built on algebraic principles, form the backbone of modern computing.

• •

CONCLUSION

Algebra, as the language of mathematics, is a powerful tool for generalizing patterns, solving equations, and modeling complex systems. Through variables, constants, and operations, it offers a symbolic framework that connects abstract concepts to real-world applications. Whether exploring linear and quadratic equations or using algebra to solve global challenges, this foundational discipline remains indispensable in science, technology, and beyond.

CHAPTER 4:
FUNDAMENTALS OF
TRIGONOMETRY

Trigonometry, the branch of mathematics that studies the relationships between angles and sides of triangles, is a cornerstone of both theoretical and applied mathematics. It has profound implications for fields as diverse as astronomy, engineering, and even music. This chapter delves into the fundamental trigonometric ratios, the unit circle, angle measurements, and explores the remarkable relevance of trigonometry in real-world contexts.

TRIGONOMETRIC RATIOS

1. Definition of Trigonometric Ratios

Trigonometric ratios describe the relationships between the sides of a right triangle relative to its angles.

- **Key Ratios**:

    1. **Sine** ($\sin$): Opposite side divided by hypotenuse:
    $$\sin\theta = \frac{\text{opposite}}{\text{hypotenuse}}$$

    2. **Cosine** ($\cos$): Adjacent side divided by hypotenuse:
    $$\cos\theta = \frac{\text{adjacent}}{\text{hypotenuse}}$$

    3. **Tangent** ($\tan$): Opposite side divided by adjacent side:
    $$\tan\theta = \frac{\text{opposite}}{\text{adjacent}}$$

- **Reciprocal Ratios**:

    1. **Cosecant** ($\csc$): Reciprocal of sine:
    $$\csc\theta = \frac{\text{hypotenuse}}{\text{opposite}}$$

    2. **Secant** ($\sec$): Reciprocal of cosine:
    $$\sec\theta = \frac{\text{hypotenuse}}{\text{adjacent}}$$

    3. **Cotangent** ($\cot$): Reciprocal of tangent:
    $$\cot\theta = \frac{\text{adjacent}}{\text{opposite}}$$

2. Understanding Ratios in Right Triangles

- Trigonometric ratios depend solely on the angles of the triangle, not the triangle's size.

- Example: For a $30°$ - $60°$ - $90°$ triangle:

$$\sin30° = \frac{1}{2}, \cos30° = \frac{\sqrt{3}}{2}, \tan30° = \frac{\sqrt{3}}{3}.$$

3. Trigonometric Identities

- Fundamental relationships between trigonometric functions: $\sin^2\theta + \cos^2\theta = 1 \quad 1 + \tan^2\theta = \sec^2\theta \quad 1 + \cot^2\theta = \csc^2\theta$

• •

THE UNIT CIRCLE AND ANGLE MEASUREMENT

1. The Unit Circle

- **Definition**: A circle with a radius of 1 centered at the origin of a Cartesian coordinate system.

- **Key Features**:

    - Points on the circle correspond to coordinates $(\cos\theta, \sin\theta)$.

    - The circle provides a visual representation of trigonometric functions.

- **Applications**:

    - Simplifies calculations of trigonometric values for standard angles.

    - Forms the basis for understanding periodic functions.

2. Angle Measurement

- **Degrees and Radians**:

    - **Degrees**: A full circle is $360°$.

    - **Radians**: A full circle is 2π radians.

- Conversion: $1 \text{ radian} = \frac{180°}{\pi}$, $1° = \frac{\pi}{180} \text{ radians}$.

- **Quadrants**:

  - The unit circle is divided into four quadrants, affecting the signs of trigonometric values:

    - Quadrant I: All positive.

    - Quadrant II: Sine positive.

    - Quadrant III: Tangent positive.

    - Quadrant IV: Cosine positive.

3. Periodicity and Symmetry

- **Periodicity**:

  - Sine and cosine repeat every 2π: $\sin(\theta + 2\pi) = \sin\theta$, $\cos(\theta + 2\pi) = \cos\theta$.

  - Tangent repeats every π: $\tan(\theta + \pi) = \tan\theta$.

- **Symmetry**:

  - **Even Functions**: $\cos(-\theta) = \cos\theta$, $\sec(-\theta) = \sec\theta$.

  - **Odd Functions**: $\sin(-\theta) = -\sin\theta$, $\tan(-\theta) = -\tan\theta$.

• •

REAL-WORLD RELEVANCE

1. Astronomy

- Trigonometry originated in ancient astronomy for calculating distances to celestial bodies.

- **Applications**:

  - Measuring the distance to stars using parallax.

  - Predicting the positions of planets and eclipses.

2. Engineering

- Trigonometric principles are used in the design of buildings, bridges, and machines.

- **Examples**:

  - Determining forces in structures (e.g., tension in cables).

  - Calculating angles in machinery and robotics.

3. Navigation

- Trigonometry enables navigation by sea, air, and space.

- **Examples**:

  - Calculating courses and bearings using spherical trigonometry.

  - GPS systems rely on trigonometric algorithms to determine positions.

4. Music

- Trigonometry models sound waves, which are periodic in nature.

- **Applications**:

    – Sine waves represent pure tones.

    – Fourier analysis breaks complex sounds into trigonometric components.

5. Physics

- Trigonometry describes oscillatory motion (e.g., pendulums, springs).

- **Applications**:

    – Calculating projectile trajectories.

    – Analyzing wave interference and diffraction.

• •

CONCLUSION

Trigonometry is an indispensable tool for understanding the relationships between angles and distances. By mastering trigonometric ratios, the unit circle, and angle measurement, we gain a deeper appreciation for its pervasive role in both theoretical and applied sciences. Whether calculating the orbits of planets or composing music, trigonometry continues to shape our understanding of the universe and its harmonious patterns.

CHAPTER 5: THE COORDINATE GEOMETRY CONNECTION

Coordinate geometry, also known as analytic geometry, unites algebra and geometry by allowing geometric shapes to be represented algebraically. This branch of mathematics provides a powerful framework for visualizing and solving problems involving points, lines, curves, and surfaces in space. By utilizing the Cartesian coordinate system, we create a bridge between the abstract world of algebra and the visual realm of geometry.

THE CARTESIAN COORDINATE SYSTEM

1. Origin and Structure

- **Definition**: A system for specifying points in a plane using ordered pairs of numbers.

- **Key Elements**:

  - **Axes**: Two perpendicular lines, the x-axis (horizontal) and y-axis (vertical), intersecting at the origin $(0,0)$.

  - **Quadrants**: The plane is divided into four regions:

    - Quadrant I: $x > 0, y > 0$

    - Quadrant II: $x < 0, y > 0$

    - Quadrant III: $x < 0, y < 0$

    - Quadrant IV: $x > 0, y < 0$

  - **Points**: Defined by coordinates (x, y), where x is the horizontal distance from the origin and y is the vertical distance.

2. Distance and Midpoint Formulas

- **Distance Formula**: The distance between two points (x_1, y_1) and (x_2, y_2) is:

$$d = \sqrt{(x_2 - x_1)^2 + (y_2 - y_1)^2}$$

- **Midpoint Formula**: The midpoint of the line segment connecting (x_1, y_1) and (x_2, y_2) is: $M = \left(\dfrac{x_1 + x_2}{2}, \dfrac{y_1 + y_2}{2} \right)$

3. Slope of a Line

- **Definition**: The slope (m) represents the steepness or inclination of a line, calculated as the ratio of vertical change (rise) to horizontal change (run): $m = \dfrac{y_2 - y_1}{x_2 - x_1}$

- **Interpretation**:

  - $m > 0$: Line slopes upward.

  - $m < 0$: Line slopes downward.

  - $m = 0$: Line is horizontal.

  - Undefined: Line is vertical.

· ·

GEOMETRIC REPRESENTATIONS

1. Equations of Lines

- **Slope-Intercept Form**: $y = mx + b$ where m is the slope and b is the y-intercept.

- **Point-Slope Form**: $y - y_1 = m(x - x_1)$ Useful for writing equations when a point and slope are known.

- **Standard Form**: $Ax + By + C = 0$ Often used for general linear equations.

2. Equations of Circles

- **Standard Form**: $(x - h)^2 + (y - k)^2 = r^2$ where (h,k) is the center and r is the radius.

- **Applications**:
  - Modeling round objects like wheels and planets.
  - Analyzing motion in circular paths.

3. Equations of Parabolas

- **Standard Form**:
  - Vertical axis: $y = ax^2 + bx + c$
  - Horizontal axis: $x = ay^2 + by + c$

- **Key Features**:
  - **Vertex**: The highest or lowest point.
  - **Axis of Symmetry**: A line dividing the parabola into two symmetric halves.
  - **Focus and Directrix**: Defines the curve as the locus of points equidistant from a point (focus) and a line (directrix).

• •

BRIDGE BETWEEN ALGEBRA AND GEOMETRY

1. Algebraic Representation of Geometric Shapes

- Coordinate geometry allows shapes to be represented by equations:

    - A line by $y = mx + b$.

    - A circle by $(x - h)^2 + (y - k)^2 = r^2$.

    - A parabola by $y = ax^2 + bx + c$.

2. Solving Geometric Problems Using Algebra

- **Intersection of Lines**: Solve simultaneous equations to find the point(s) of intersection.

- **Distance Between Shapes**: Use algebraic formulas to calculate distances, such as between a point and a line:

$$d = \frac{|Ax_1 + By_1 + C|}{\sqrt{A^2 + B^2}}$$

- **Optimization Problems**: Algebraic techniques are used to optimize dimensions of shapes, such as maximizing area or minimizing perimeter.

3. Visualizing Algebraic Equations

- Graphing equations provides an intuitive understanding of

algebraic relationships.

- Example:

  - The equation $y = 2x + 3$ represents a line with slope 2 and y-intercept 3.

  - The equation $x^2 + y^2 = 25$ represents a circle centered at the origin with radius 5.

· ·

APPLICATIONS
IN REAL LIFE

1. Engineering and Architecture

- Coordinate geometry is used to design and analyze structures.

- Example: Calculating stress on beams and angles in trusses.

2. Physics and Astronomy

- Modeling trajectories of objects using parabolas and circles.

- Example: The path of a projectile or the orbit of a planet.

3. Computer Graphics

- Coordinate geometry forms the basis for 2D and 3D modeling.

- Example: Rendering shapes and animations in video games.

4. Navigation

- GPS systems rely on coordinate geometry to pinpoint locations on Earth.

• •

CONCLUSION

Coordinate geometry serves as a vital bridge between algebra and geometry, enabling the visualization and solution of complex problems. By combining the precision of algebraic equations with the clarity of geometric representations, this discipline enhances our ability to model and analyze the world around us. From understanding the properties of lines, circles, and parabolas to solving real-world challenges, coordinate geometry underscores the interconnected nature of mathematical thought.

CHAPTER 6:
GEOMETRY MEETS
TRIGONOMETRY

• •

Geometry and trigonometry are deeply intertwined branches of mathematics that complement each other in understanding shapes, angles, and distances. This chapter explores how trigonometric principles apply to triangles and circles, introduces the Laws of Sines and Cosines, and delves into their applications in fields like engineering and navigation.

• •

TRIANGLES AND THE LAWS OF SINES AND COSINES

1. Types of Triangles in Geometry

- **Right Triangle**: Contains one $90°$ angle.

- **Acute Triangle**: All angles are less than $90°$.

- **Obtuse Triangle**: Contains one angle greater than $90°$.

- **Equilateral Triangle**: All sides and angles are equal.

- **Isosceles Triangle**: Two sides are equal in length.

- **Scalene Triangle**: All sides and angles are different.

2. Law of Sines

The Law of Sines relates the ratios of the lengths of sides of a triangle to the sines of their opposite angles.

$$\frac{a}{\sin A} = \frac{b}{\sin B} = \frac{c}{\sin C}$$

- **Where:**

  - a, b, c: Lengths of the sides.

  - A, B, C: Opposite angles.

- **Applicability:**

–   Useful for solving non-right triangles in the **ASA (Angle-Side-Angle)** or **AAS (Angle-Angle-Side)** cases.

- **Example**: In a triangle with $A = 40^\circ$, $B = 60^\circ$, and $a = 10$:

$$\frac{10}{\sin 40^\circ} = \frac{b}{\sin 60^\circ}$$ Solve for b: $$b = \frac{10 \cdot \sin 60^\circ}{\sin 40^\circ}.$$

3. Law of Cosines

The Law of Cosines generalizes the Pythagorean theorem to all triangles, relating the lengths of sides to the cosine of their included angle.

$$c^2 = a^2 + b^2 - 2ab \cdot \cos C$$

- **Where**:

    – a, b, c: Lengths of the sides.

    – C: Angle opposite side c.

- **Applicability**:

    –   Solves triangles in the **SAS (Side-Angle-Side)** or **SSS (Side-Side-Side)** cases.

- **Example**: In a triangle with $a = 5$, $b = 7$, and $C = 60^\circ$:

$$c^2 = 5^2 + 7^2 - 2(5)(7)\cos 60^\circ$$ Solve for c: $c = \sqrt{74}$.

• •

CIRCLES AND ANGLES

1. Central Angles

- **Definition**: An angle whose vertex is at the center of the circle.

- **Properties**:

  - The measure of a central angle equals the measure of the arc it subtends.

  - Example: A central angle of 90° subtends a quarter of the circle's circumference.

2. Inscribed Angles

- **Definition**: An angle whose vertex lies on the circle, with sides that intersect the circle.

- **Properties**:

  - The measure of an inscribed angle is half the measure of the arc it subtends.

  - Example: An inscribed angle subtending a semicircle is always 90°.

3. Relationships Between Central and Inscribed Angles

- **Key Property**:

  - If an inscribed angle and a central angle subtend the same arc, the inscribed angle is half the central angle.

4. Applications in Trigonometry

- Trigonometric functions ($\sin, \cos, \tan$) can be used to calculate angles and arc lengths in circles.

- **Arc Length**: $L = r \cdot \theta$

    - Where r is the radius and θ is the angle in radians.

- **Sector Area**: $A = \frac{1}{2} r^2 \theta$

. . . .

APPLICATIONS IN ENGINEERING AND NAVIGATION

1. Engineering

Trigonometry and geometry are essential for designing and analyzing structures and systems.

- **Structural Engineering**:

    - Determining load distributions and forces in trusses and bridges.

    - Example: Using the Law of Cosines to calculate stress in beams.

- **Mechanical Engineering**:

    - Calculating angular motion in rotating systems.

    - Designing gears and pulleys based on angular relationships.

- **Electrical Engineering**:

    - Using sine and cosine functions to model alternating current (AC) waveforms.

    - Example: Voltage and current waveforms follow trigonometric patterns.

2. Navigation

Trigonometry plays a critical role in determining position and direction in navigation.

- **Celestial Navigation**:
    - Mariners and aviators use spherical trigonometry to calculate positions based on celestial bodies.
    - Example: Using the Law of Sines to determine latitude from star altitudes.

- **GPS Systems**:
    - GPS algorithms use coordinate geometry and trigonometry to calculate distances between satellites and receivers.
    - Example: Triangulation relies on the properties of triangles to pinpoint locations.

- **Aviation**:
    - Pilots use trigonometry to determine angles of ascent and descent.
    - Example: Calculating heading adjustments due to wind using vector trigonometry.

3. Surveying and Cartography
- Trigonometry is used to measure distances and angles for mapping land and constructing large projects.
- Example:
    - A surveyor uses the Law of Sines to calculate the distance between two points on uneven terrain.

4. Sound and Wave Propagation
- Trigonometry models wave phenomena, such as sound and light.
- Example:
    - Engineers calculate the diffraction of waves around obstacles using trigonometric principles.

• •

CONCLUSION

The fusion of geometry and trigonometry provides powerful tools for solving problems involving triangles and circles. The Laws of Sines and Cosines extend the applicability of trigonometry to non-right triangles, while concepts like central and inscribed angles bridge the gap between linear and circular geometry. These principles have far-reaching applications in engineering, navigation, and countless other fields, underscoring their essential role in understanding and shaping the world around us.

CHAPTER 7: PROOFS AND THEOREMS IN GEOMETRY

• •

Proofs and theorems form the backbone of geometry, providing logical explanations for the relationships between shapes, angles, and lines. This chapter explores core geometric proofs, such as triangle congruence and the Pythagorean theorem, introduces algebraic approaches to proofs, and highlights the elegance of logical reasoning in mathematical thought.

• •

CORE PROOFS IN GEOMETRY

1. Triangle Congruence Proofs

Triangle congruence establishes when two triangles are identical in shape and size, based on their sides and angles.

- **Key Congruence Criteria:**
  1. **Side-Side-Side (SSS):**
     - If all three sides of one triangle are equal to the corresponding sides of another triangle, the triangles are congruent.
     - **Proof:**
       - Place the triangles in a Cartesian plane.
       - Use the distance formula to show that corresponding sides are equal.
  2. **Side-Angle-Side (SAS):**
     - If two sides and the included angle of one triangle are equal to the corresponding parts of another triangle, the triangles are congruent.
     - **Proof:**
       - Superimpose one triangle on the other using the included angle as a pivot.
       - Demonstrate that the third side aligns, proving congruence.
  3. **Angle-Side-Angle (ASA):**

- If two angles and the included side of one triangle are equal to the corresponding parts of another triangle, the triangles are congruent.

- **Proof:**

    - Use the sum of angles in a triangle ($180°$) to confirm that the third angle is also equal.

    - Show that all sides align.

4. **Hypotenuse-Leg (HL)** (specific to right triangles):

- If the hypotenuse and one leg of a right triangle are equal to the corresponding parts of another triangle, the triangles are congruent.

- **Proof:**

    - Use the Pythagorean theorem to confirm equality of the other leg.

2. Pythagorean Theorem

The Pythagorean theorem relates the lengths of the sides of a right triangle: $a^2 + b^2 = c^2$

- **Proofs:**

    1. **Geometric Proof:**

    - Construct squares on each side of the triangle.

    - Show that the areas of the smaller squares sum to the area of the largest square.

    2. **Algebraic Proof:**

    - Divide the hypotenuse square into two smaller triangles.

    - Use algebra to demonstrate that the sum of their areas equals the area of the hypotenuse square.

3. Angle Sum Theorem

The sum of the interior angles of a triangle is always $180°$.

- **Proof**:

  – Extend one side of the triangle.

  – Use alternate interior angles and linear pairs to show that the angles add to $180°$.

4. Midpoint Theorem

The line segment joining the midpoints of two sides of a triangle is parallel to the third side and half as long.

- **Proof**:

  – Place the triangle in a coordinate system.

  – Use the midpoint formula and slope calculations to prove the parallelism and proportionality.

• •

ALGEBRAIC APPROACHES TO GEOMETRIC PROOFS

1. Using Coordinates in Proofs

Coordinate geometry allows geometric properties to be proven algebraically.

- **Example**: Proving the Pythagorean theorem.

  - Place a right triangle with vertices at $(0,0)$, $(a,0)$, and $(0,b)$.

  - Use the distance formula to calculate the hypotenuse: $c = \sqrt{a^2 + b^2}$ Squaring both sides yields $a^2 + b^2 = c^2$.

2. Vector Methods

Vectors provide a powerful tool for proving geometric relationships.

- **Example**: Proving that the diagonals of a parallelogram bisect each other.

  - Represent the parallelogram with vertices A, B, C, D.

  - Use vector addition to find the midpoint of each diagonal and show they coincide.

3. Algebraic Representation of Shapes

- Represent geometric figures with equations.

TONY YUSTEIN

- **Example**: Proving the equation of a circle.

  – Use the definition of a circle as the set of points equidistant from a center (h,k): $(x - h)^2 + (y - k)^2 = r^2$

· ·

UNDERSTANDING THE BEAUTY OF LOGICAL REASONING

1. The Structure of a Proof

A proof is a logical argument that establishes the truth of a statement beyond doubt.

- **Steps in a Proof**:
    1. **Given**: State the problem or conditions.
    2. **To Prove**: Specify the statement to be demonstrated.
    3. **Diagram**: Draw a figure (if applicable) to clarify the problem.
    4. **Proof**: Provide a step-by-step logical explanation using definitions, axioms, and previously proven theorems.

2. Axioms and Postulates

- Geometry builds upon self-evident truths, such as Euclid's postulates.
- **Example**: Parallel Postulate.
    - Through a point not on a line, exactly one parallel line can be drawn.

3. Why Proofs Matter

Proofs ensure rigor and consistency in mathematics,

distinguishing true statements from conjectures.

- **Intellectual Beauty**:

    –   Proofs reveal unexpected connections, like the Pythagorean theorem's relationship to algebra and geometry.

- **Practical Applications**:

    –   Proofs validate techniques used in engineering, physics, and computer science.

4. Creativity in Proofs

- Some theorems, like the Pythagorean theorem, have multiple proofs ranging from geometric to algebraic.

- This diversity highlights the richness of mathematical thought.

• •

APPLICATIONS OF PROOFS IN THE REAL WORLD

1. Engineering

- Proofs of structural stability ensure safe bridge and building designs.

- Example:

  – The properties of triangles are used to calculate forces in trusses.

2. Computer Science

- Geometric proofs underpin algorithms for graphics rendering and machine learning.

- Example:

  – Proving the correctness of shortest-path algorithms using coordinate geometry.

3. Physics

- Proofs validate the equations governing motion and forces.

- Example:

  – Deriving the trajectory of a projectile using vector methods.

4. Cryptography

- Algebraic proofs are used to ensure the security of encryption systems.

CONCLUSION

Proofs are the cornerstone of geometry, blending logic, algebra, and creativity to unveil the hidden truths of shapes and spaces. From foundational theorems like triangle congruence and the Pythagorean theorem to modern applications in engineering and cryptography, proofs exemplify the elegance and utility of logical reasoning. As tools of rigorous thought, they not only validate mathematical principles but also illuminate the inherent beauty of the subject.

CHAPTER 8: VECTORS IN ALGEBRA AND GEOMETRY

• •

Vectors are a fundamental concept in mathematics that bridge algebra and geometry, representing quantities with both magnitude and direction. They are indispensable tools for solving problems in physics, engineering, and computer graphics. This chapter introduces the concept of vectors, explains their representation in two and three dimensions, and explores their wide-ranging applications.

• •

INTRODUCTION
TO VECTORS

1. Definition of a Vector

A vector is a mathematical object that has both **magnitude** (size) and **direction**. Unlike scalars, which are quantities described by magnitude alone (e.g., mass or temperature), vectors include direction, making them ideal for representing quantities like force, velocity, and displacement.

- **Example**: A car moving at 50 km/h north is described by a vector, as it has both speed (magnitude) and direction (north).

2. Notation and Representation

- Vectors are often denoted by boldface letters (v) or letters with an arrow above them ($\vec{v}$).

- A vector is represented geometrically as an arrow:

    - The length of the arrow represents the magnitude.

    - The direction of the arrow represents the direction of the vector.

3. Magnitude of a Vector

The magnitude of a vector v is denoted by $|v|$ or $\|v\|$. For a vector in 2D space with components (x, y), the magnitude is calculated as: $|v| = \sqrt{x^2 + y^2}$

4. Unit Vectors

A unit vector has a magnitude of 1 and is used to indicate direction. A vector v can be converted into a unit vector $\hat{v}$ as follows: $\hat{v} = \dfrac{v}{|v|}$

· ·

REPRESENTATION IN 2D AND 3D SPACES

1. Vectors in 2D Space

- A vector in two dimensions is represented as an ordered pair of components: $v = (x,y)$

  - x: The horizontal component.

  - y: The vertical component.

- **Vector Addition and Subtraction**:

  - To add or subtract two vectors $u = (u_x, u_y)$ and $v = (v_x, v_y)$:
  $$u + v = (u_x + v_x, u_y + v_y) \quad u - v = (u_x - v_x, u_y - v_y)$$

- **Scalar Multiplication**:

  - Multiplying a vector by a scalar k: $kv = (kx, ky)$

2. Vectors in 3D Space

- A vector in three dimensions is represented as an ordered triple: $v = (x,y,z)$

  - x, y, z: Components along the three axes.

- **Magnitude in 3D**: $|v| = \sqrt{x^2 + y^2 + z^2}$

- **Cross Product**: The cross product of two vectors u and v

$$u \times v = \begin{vmatrix} i & j & k \\ u_x & u_y & u_z \\ v_x & v_y & v_z \end{vmatrix}$$

produces a vector perpendicular to both:

- Where i, j, k are the unit vectors along the x-, y-, and z-axes.

- **Dot Product**: The dot product of two vectors measures their relative orientation: $u \cdot v = u_x v_x + u_y v_y + u_z v_z$

  - Used to find the angle between two vectors: $\cos\theta = \dfrac{u \cdot v}{|u||v|}$

. . . .

APPLICATIONS IN PHYSICS, ENGINEERING, AND COMPUTER GRAPHICS

1. Physics

- **Force and Motion**:

    – Forces acting on an object are represented as vectors, allowing the calculation of net force through vector addition.

    – Example: Combining gravitational force and normal force in an inclined plane problem.

- **Projectile Motion**:

    – The motion of projectiles is analyzed by decomposing velocity into horizontal and vertical components using vectors.

- **Electric and Magnetic Fields**:

    – Field intensities at a point are expressed as vectors, aiding in visualizing interactions between charges or currents.

2. Engineering

- **Structural Engineering**:

    – Vectors are used to analyze forces and stresses in

structures like bridges and buildings.

- – Example: Resolving forces in a truss to ensure stability.
- **Mechanical Engineering**:
  - – Calculating torque involves the cross product of force and distance vectors: $\tau = r \times F$
- **Fluid Dynamics**:
  - – Representing fluid flow direction and magnitude at different points in space.

3. Computer Graphics

- **3D Rendering**:
  - – Vectors are used to define positions, directions, and surface normals in 3D models.
  - – Example: Lighting calculations in 3D rendering use the dot product to determine brightness based on the angle of incidence.
- **Animation**:
  - – Motion paths of objects are described using vector equations.
- **Collision Detection**:
  - – Vectors help in calculating the direction and magnitude of movement during collisions in video games or simulations.

4. Navigation and Robotics

- **Navigation**:
  - – Vectors are used in GPS systems to calculate distances and directions.
- **Robotics**:
  - – Motion planning for robotic arms relies on vector

mathematics to calculate positions in 3D space.

• •

VISUALIZING VECTORS

- **Vector Fields**:

  – Used to represent quantities that vary across space, like wind velocity or magnetic fields.

  – Example: Meteorologists use vector fields to model wind patterns.

- **Graphical Representation**:

  – Vectors can be added, subtracted, or multiplied visually using the triangle or parallelogram methods.

• •

CONCLUSION

Vectors are a versatile tool that seamlessly connect algebra and geometry, providing a robust framework for analyzing physical phenomena, designing structures, and creating virtual worlds. By understanding their magnitude, direction, and applications in 2D and 3D spaces, we unlock a powerful mathematical language with limitless possibilities in physics, engineering, computer graphics, and beyond.

CHAPTER 9: TRANSFORMATIONS AND SYMMETRY

• •

Transformations and symmetry are essential concepts in geometry that describe how shapes move, change, and remain balanced. Transformations include translations, rotations, reflections, and dilations, each altering a shape's position, orientation, or size. Symmetry reveals the beauty of balance and order in geometry, as seen in nature, art, and architecture. This chapter delves into the mathematical principles of transformations, their algebraic representation using matrices, and the universal presence of symmetry.

• •

GEOMETRIC TRANSFORMATIONS

Geometric transformations describe how figures change or move while retaining certain properties.

1. Translation

- **Definition**: Moving a shape without changing its size, shape, or orientation.

- **Properties**:

  - Every point of the shape moves the same distance in the same direction.

  - Preserves congruence (shape and size remain unchanged).

- **Representation**:

  - A point (x,y) is translated by (h,k): $(x,y) \rightarrow (x + h, y + k)$

2. Rotation

- **Definition**: Turning a shape around a fixed point, known as the center of rotation.

- **Properties**:

  - The shape's size and angles remain unchanged.

  - Points rotate through a specified angle (θ).

- **Representation**:

  - Rotating a point (x,y) around the origin:

$$(x,y) \rightarrow (x\cos\theta - y\sin\theta, x\sin\theta + y\cos\theta)$$

- For clockwise rotation, negate θ.

3. Reflection

- **Definition**: Flipping a shape over a line, called the line of reflection, creating a mirror image.

- **Properties**:

  - Shape and size remain unchanged.

  - The reflected shape is congruent but reversed.

- **Representation**:

  - Reflecting a point (x,y) over:

    - The x-axis: $(x,y) \rightarrow (x, -y)$

    - The y-axis: $(x,y) \rightarrow (-x, y)$

    - The line $y = x$: $(x,y) \rightarrow (y,x)$

4. Dilation

- **Definition**: Resizing a shape by scaling it larger or smaller while maintaining proportionality.

- **Properties**:

  - Shapes remain similar (same angles, proportional sides).

  - Requires a center of dilation and a scale factor (k).

- **Representation**:

  - A point (x,y) is dilated with respect to the origin: $(x,y) \rightarrow (kx, ky)$

ALGEBRAIC REPRESENTATION USING MATRICES

Matrices provide a powerful tool for representing and performing geometric transformations, especially in higher dimensions and computer graphics.

1. Translation

- Represented by adding a translation vector:

$$T = \begin{bmatrix} x + h \\ y + k \end{bmatrix}$$

2. Rotation

- Represented by a rotation matrix:

$$R = \begin{bmatrix} \cos\theta & -\sin\theta \\ \sin\theta & \cos\theta \end{bmatrix}$$

  - Apply the matrix to a vector $v = \begin{bmatrix} x \\ y \end{bmatrix}$ to find the rotated coordinates:

$$R \cdot v = \begin{bmatrix} x\cos\theta - y\sin\theta \\ x\sin\theta + y\cos\theta \end{bmatrix}$$

3. Reflection

- Reflection matrices for common axes:

  - Over the x-axis:

$$M_x = \begin{bmatrix} 1 & 0 \\ 0 & -1 \end{bmatrix}$$

$$M_y = \begin{bmatrix} -1 & 0 \\ 0 & 1 \end{bmatrix}$$

- Over the y-axis:

$$M_{y=x} = \begin{bmatrix} 0 & 1 \\ 1 & 0 \end{bmatrix}$$

- Over $y = x$:

4. Dilation

$$D = \begin{bmatrix} k & 0 \\ 0 & k \end{bmatrix}$$

• Dilation with scale factor k:

5. Combined Transformations

• Multiple transformations can be applied sequentially by multiplying their corresponding matrices.

• •

SYMMETRY IN NATURE, ART, AND ARCHITECTURE

Symmetry is a fundamental aspect of aesthetics and structural integrity, characterized by balance and proportionality.

1. Symmetry in Nature

- **Radial Symmetry**: Found in flowers, starfish, and snowflakes, where symmetry radiates from a central point.

- **Bilateral Symmetry**: Seen in animals and humans, where one side mirrors the other.

- **Fractals**: Patterns that repeat at different scales, such as ferns and coastlines.

2. Symmetry in Art

- Symmetry is a cornerstone of artistic design, conveying harmony and balance.

- **Examples**:

  - The intricate tiling in Islamic art demonstrates reflection and rotational symmetry.

  - Renaissance paintings, like Leonardo da Vinci's *Vitruvian Man*, emphasize bilateral symmetry.

3. Symmetry in Architecture

- Symmetry enhances structural stability and aesthetic appeal in architecture.

- **Examples**:

  – The Taj Mahal exhibits perfect bilateral symmetry.

  – Gothic cathedrals often use reflectional and rotational symmetry in their rose windows.

4. Mathematical Analysis of Symmetry

- Symmetry groups in mathematics classify patterns and transformations that preserve symmetry.

- **Example**:

  – The group of rotational symmetries of a square includes rotations by $0°$, $90°$, $180°$, and $270°$.

• •

APPLICATIONS OF TRANSFORMATIONS AND SYMMETRY

1. Computer Graphics

- Transformations are essential for rendering objects in video games and animations.

- Symmetry is used to optimize designs and reduce computational requirements.

2. Engineering

- Symmetry in bridge and building designs ensures even weight distribution and structural stability.

- Transformations model forces acting on objects.

3. Robotics

- Transformations guide robotic arms for precise movement and positioning.

4. Physics

- Symmetry principles simplify equations in quantum mechanics and particle physics.

• •

CONCLUSION

Transformations and symmetry form the heart of geometry, bridging algebraic precision with visual elegance. Through translations, rotations, reflections, and dilations, we understand how shapes move and change, while symmetry reveals balance and order in the natural and constructed world. Together, they provide tools for innovation across art, engineering, and technology, showcasing the timeless relevance of mathematics.

CHAPTER 10:
PERIODIC PHENOMENA AND TRIGONOMETRIC FUNCTIONS

• •

Periodic phenomena are natural occurrences that repeat over time, such as the oscillation of waves, rotation of planets, or the rhythm of a heartbeat. Trigonometric functions like sine and cosine are the mathematical tools that describe these patterns. This chapter delves into the exploration of sine and cosine waves, their role in modeling real-world periodic phenomena, and their advanced applications in Fourier series and signal processing.

• •

EXPLORING SINE AND COSINE WAVES

1. Introduction to Periodic Functions

- **Definition**: A function is periodic if it repeats its values at regular intervals, known as the period.

- **Mathematical Representation**: $f(x + T) = f(x)$ for all x, where T is the period.

2. The Sine Function

- **Equation**: $y = A\sin(\omega x + \phi)$

  - A: Amplitude (height of the wave).

  - ω: Angular frequency ($\omega = \dfrac{2\pi}{T}$).

  - ϕ: Phase shift (horizontal displacement).

  - T: Period ($T = \dfrac{2\pi}{\omega}$).

- **Graphical Features**:

  - Starts at $y = 0$ when $x = 0$ (if $\phi = 0$).

  - Oscillates between $-A$ and $+A$.

  - Repeats every 2π for $\omega = 1$.

3. The Cosine Function

- **Equation**: $y = A\cos(\omega x + \phi)$

    - Shares properties with the sine function but starts at $y = A$ when $x = 0$ (if $\phi = 0$).

- **Graphical Features**:

    - Similar to the sine function but shifted by $\frac{\pi}{2}$:

$$\cos(x) = \sin\left(x + \frac{\pi}{2}\right)$$

4. Properties of Sine and Cosine Waves

- **Periodicity**:

    - Both functions repeat every 2π radians (or 360°) when $\omega = 1$.

- **Symmetry**:

    - Sine is an odd function: $\sin(-x) = -\sin(x)$.

    - Cosine is an even function: $\cos(-x) = \cos(x)$.

- **Applications**:

    - Describe oscillations in mechanical and electromagnetic systems.

• •

MODELING REAL-WORLD PERIODIC PATTERNS

1. Sound Waves

- **Nature of Sound**:

  - Sound is a longitudinal wave, where air molecules oscillate to produce pressure variations.

- **Mathematical Representation**:

  - Sound waves can be modeled as sine waves: $y(t) = A\sin(2\pi ft + \phi)$, where f is the frequency in hertz (Hz).

- **Example**:

  - A tuning fork vibrating at 440 Hz produces the note A, modeled as: $y(t) = A\sin(2\pi \cdot 440 \cdot t)$.

2. Light Waves

- **Nature of Light**:

  - Light is an electromagnetic wave, with oscillating electric and magnetic fields.

- **Mathematical Representation**:

  - Light intensity can be modeled using sine and cosine functions: $E(t) = E_0\cos(\omega t)$, where E_0 is the amplitude of the electric field.

- **Applications**:
    - Modeling interference patterns in optics.
    - Describing polarization and diffraction.

3. Mechanical Oscillations
- **Pendulums and Springs**:
    - The displacement of a pendulum or a mass-spring system follows a sinusoidal pattern.
- **Example**:
    - The position of a mass on a spring is given by: $x(t) = A\cos(\omega t + \phi)$.

4. Biological Rhythms
- **Heartbeats and Circadian Rhythms**:
    - Periodic biological processes, like heart rates and sleep cycles, can be analyzed using sine waves.
- **Example**:
    - A heartbeat signal can be approximated by a combination of sinusoidal functions.

• •

FOURIER SERIES AND SIGNAL PROCESSING

1. Fourier Series
- **Definition**:

  - Any periodic function can be represented as an infinite sum of sine and cosine functions:

  $f(x) = a_0 + \sum_{n=1}^{\infty} (a_n \cos(nx) + b_n \sin(nx))$.

- **Coefficients**:

  - a_0: Represents the average value (DC component).

  - a_n and b_n: Represent the amplitudes of cosine and sine components.

- **Applications**:

  - Fourier series decompose complex periodic signals into simple harmonics.

2. Signal Processing
- **Definition**:

  - Signal processing involves analyzing, modifying, and synthesizing signals using mathematical techniques.

- **Role of Trigonometry**:

  - Sine and cosine waves form the basis of signal representation.

- **Applications**:

  - **Audio Engineering**:

    - Removing noise or enhancing frequencies in music.

  - **Image Compression**:

    - JPEG compression uses Fourier transforms to represent images efficiently.

  - **Telecommunications**:

    - Modulation of signals for data transmission.

3. Fourier Transform

- **Extension of Fourier Series**:

  - The Fourier transform extends the Fourier series to non-periodic functions.

- **Formula**: $F(\omega) = \int_{-\infty}^{\infty} f(t)e^{-i\omega t} dt,$ where $F(\omega)$ represents the frequency spectrum of the signal.

- **Applications**:

  - Analyzing frequency components in speech, music, and seismic data.

• •

APPLICATIONS OF PERIODIC PHENOMENA

1. Engineering

- **Structural Analysis**:

  - Oscillations in bridges and buildings are modeled with sine and cosine functions.

- **Electronics**:

  - AC circuits follow sinusoidal voltage and current patterns.

2. Medicine

- **Heart Rate Monitoring**:

  - ECG signals are analyzed using Fourier techniques to detect abnormalities.

- **Brain Waves**:

  - EEG signals are modeled as periodic waves to study neurological disorders.

3. Astronomy

- **Planetary Motion**:

  - Trigonometric functions model the periodic orbits of planets and moons.

- **Wave Analysis**:

-     Radio signals from celestial objects are processed using Fourier transforms.

· ·

CONCLUSION

Trigonometric functions like sine and cosine are fundamental to understanding periodic phenomena in nature and technology. From modeling sound and light waves to analyzing complex signals using Fourier series, these tools reveal the underlying patterns of oscillatory systems. The versatility of trigonometric functions highlights their importance in bridging mathematics with real-world applications, enabling advances in engineering, medicine, and beyond.

CHAPTER 11: CONIC SECTIONS AND THEIR APPLICATIONS

Conic sections are the curves formed when a plane intersects a double-napped cone. These elegant shapes—parabolas, ellipses, and hyperbolas—are foundational to both pure and applied mathematics. Their properties and equations provide powerful tools for understanding phenomena in astronomy, satellite design, physics, and mechanics. This chapter explores the algebraic and geometric perspectives of conic sections and their applications in real-world scenarios.

THE GEOMETRY OF CONIC SECTIONS

1. Definition and Formation

Conic sections are created by the intersection of a plane with a double-napped cone. Depending on the angle and position of the plane relative to the cone, different curves are produced:

- **Parabola**: Formed when the plane is parallel to a generating line of the cone.

- **Ellipse**: Formed when the plane intersects the cone at an angle less than that of the generating line but does not pass through the cone's base.

- **Hyperbola**: Formed when the plane intersects both nappes of the cone.

2. Geometric Properties

- **Parabola**:

  – A parabola is the set of all points equidistant from a fixed point (focus) and a fixed line (directrix).

  – Symmetric about its axis of symmetry.

- **Ellipse**:

  – An ellipse is the set of all points for which the sum of the distances to two fixed points (foci) is constant.

  – Symmetric about both axes.

- **Hyperbola**:

  – A hyperbola is the set of all points where the absolute

difference of the distances to two fixed points (foci) is constant.

– Comprises two disconnected curves (branches).

3. Components of Conic Sections

- **Focus**: The fixed point(s) defining the conic.

- **Directrix**: The fixed line(s) defining the conic.

- **Axis of Symmetry**: A line passing through the focus and perpendicular to the directrix.

- **Vertices**: Points of intersection of the conic with its axis of symmetry.

• •

ALGEBRAIC REPRESENTATIONS OF CONIC SECTIONS

1. Standard Equations

- **Parabola:**

  - Vertical axis: $y = ax^2 + bx + c$ or $(x - h)^2 = 4p(y - k)$

  - Horizontal axis: $(y - k)^2 = 4p(x - h)$

- **Ellipse:**

  - Center at (h,k): $\dfrac{(x - h)^2}{a^2} + \dfrac{(y - k)^2}{b^2} = 1$

    - a: Semi-major axis, b: Semi-minor axis, c: Distance from center to foci ($c^2 = a^2 - b^2$).

- **Hyperbola:**

  - Center at (h,k): $\dfrac{(x - h)^2}{a^2} - \dfrac{(y - k)^2}{b^2} = 1$

    - a: Distance from center to vertices, b: Distance related to the slopes of the asymptotes, c: Distance from center to foci ($c^2 = a^2 + b^2$).

2. Graphical Interpretation

- Plotting the equations of conic sections reveals their shape and key features, such as vertices, foci, and asymptotes (for hyperbolas).

• •

APPLICATIONS IN ASTRONOMY AND SATELLITE DESIGN

1. Parabolas

- **Reflection Properties**:

  - Parabolas reflect parallel rays to a single focus, making them ideal for designing reflective surfaces.

- **Applications**:

  - **Telescopes**: Parabolic mirrors focus light from distant stars to create clear images.

  - **Satellite Dishes**: Parabolic reflectors focus incoming signals at the receiver located at the focus.

2. Ellipses

- **Orbital Mechanics**:

  - The orbits of planets, moons, and satellites are elliptical, with one focus occupied by the central body (e.g., the Sun for planets, Earth for satellites).

  - **Kepler's First Law of Planetary Motion**:

    - Planets move in elliptical orbits with the Sun at one focus.

- **Applications**:

  - Designing stable satellite orbits for communication and navigation systems.

    –   Analyzing the trajectories of comets and asteroids.

3. Hyperbolas

- **Wave Propagation**:

  –   Hyperbolic shapes are used to model wave fronts, such as sound waves or light waves in certain mediums.

- **Applications**:

  –   **Radio Astronomy**: Hyperbolic reflectors collect and focus electromagnetic waves from space.

  –   **Spacecraft Trajectories**:

      •   Hyperbolic trajectories describe the paths of spacecraft escaping a planet's gravitational pull.

• •

CONNECTION TO ADVANCED PHYSICS AND MECHANICS

1. Newtonian Mechanics and Conic Sections

- **Gravitational Orbits**:

    - Conic sections describe the possible trajectories of objects under gravity:

        - **Elliptical**: Bound orbits.

        - **Parabolic**: Escape trajectories at exactly the escape velocity.

        - **Hyperbolic**: Escape trajectories exceeding the escape velocity.

- **Energy Considerations**:

    - Total energy determines the conic section:

        - $E < 0$: Ellipse (bound orbit).

        - $E = 0$: Parabola (escape orbit).

        - $E > 0$: Hyperbola (escape with excess energy).

2. Relativity and Ellipses

- Elliptical paths are critical in general relativity for modeling orbits in curved spacetime.

- Example:

–   The orbit of Mercury, influenced by the curvature of spacetime, deviates slightly from a perfect ellipse (precession).

3. Engineering and Parabolic Motion

•   **Projectile Motion**:

–   Parabolic paths describe the motion of projectiles under gravity in the absence of air resistance.

–   Example:

•   Calculating the trajectory of a ball, bullet, or rocket.

• •

REAL-WORLD APPLICATIONS

1. Engineering

- **Bridge Design**:

    - Parabolic arches are used in bridges for their strength and aesthetic appeal.

- **Optics**:

    - Elliptical and parabolic lenses focus light for various optical instruments.

2. Communication

- **Satellite Design**:

    - Elliptical orbits optimize the placement of geostationary satellites for global coverage.

3. Transportation

- **Highways and Overpasses**:

    - Hyperbolic and parabolic curves are used in highway design to ensure smooth transitions and structural integrity.

• •

CONCLUSION

Conic sections—parabolas, ellipses, and hyperbolas—are more than mathematical curiosities; they are fundamental to understanding and shaping the world around us. From modeling planetary orbits and designing satellite trajectories to engineering reflective surfaces and analyzing wave propagation, these shapes showcase the profound connection between algebra, geometry, and the physical universe. As bridges between abstract mathematics and practical applications, conic sections remain indispensable in advancing technology and science.

CHAPTER 12:
OPTIMIZATION
PROBLEMS

• •

Optimization is the mathematical process of finding the best solution to a problem, often by maximizing or minimizing a particular quantity. Whether minimizing distances, maximizing areas, or improving efficiency, optimization is crucial in numerous fields, including economics, biology, and logistics. This chapter explores real-world optimization problems, highlights applications across disciplines, and introduces calculus as a tool to enhance problem-solving.

• •

THE BASICS OF OPTIMIZATION

1. What is Optimization?

- **Definition**: Optimization involves finding the maximum or minimum value of a function within a defined domain.

- **Key Concepts**:

  - **Objective Function**: The function to be optimized (e.g., profit, cost, area).

  - **Constraints**: Conditions or limits on the variables (e.g., budget, resources).

- **Types of Optimization**:

  - **Unconstrained**: No restrictions on the variables.

  - **Constrained**: Variables must satisfy certain conditions.

2. Optimization in Geometry

- Optimization often involves geometric properties such as distances, areas, and volumes.

- **Example**:

  - Finding the dimensions of a rectangle with maximum area given a fixed perimeter.

• •

REAL-WORLD EXAMPLES OF OPTIMIZATION

1. Minimizing Distances

- **Problem**:

    – Find the shortest path between two points.

- **Example**:

    – A delivery driver wants to minimize travel time by taking the shortest route.

- **Mathematical Approach**:

    – Use the distance formula: $d = \sqrt{(x_2 - x_1)^2 + (y_2 - y_1)^2}$

    – Apply constraints, such as obstacles or specific waypoints.

2. Maximizing Areas

- **Problem**:

    – Design a fence to enclose the maximum area with a fixed length of fencing material.

- **Example**:

    – A farmer wants to maximize the area of a rectangular field with 100 meters of fencing.

- **Solution**:

-     Let the length be l and the width be w, with the constraint: $2l + 2w = 100$

-     Maximize the area $A = l \cdot w$ using substitution and calculus.

3. Minimizing Costs

- **Problem**:

  - Minimize production costs while meeting demand.

- **Example**:

  - A factory wants to minimize the cost of manufacturing 1,000 units while considering material and labor expenses.

- **Solution**:

  - Develop a cost function $C(x)$, where x represents the number of units produced.

  - Use calculus to find the value of x that minimizes $C(x)$.

• •

APPLICATIONS ACROSS DISCIPLINES

1. Economics

- **Profit Maximization**:

    – Businesses aim to maximize profit by balancing revenue and costs.

    – **Example**:

        • A company selling a product at price p has revenue $R(p) = p \cdot q(p)$, where $q(p)$ is the quantity demanded.

        • Profit is given by: $P(p) = R(p) - C(p)$, where $C(p)$ is the cost function.

        • Use calculus to find the price p that maximizes $P(p)$.

- **Resource Allocation**:

    – Optimization helps allocate resources efficiently in production, marketing, and logistics.

2. Biology

- **Animal Behavior**:

    – Animals optimize energy expenditure when foraging for food.

    – **Example**:

- Birds minimize the energy spent flying to a feeding location by choosing the shortest route.

- **Population Dynamics**:

    - Models like the logistic growth equation optimize understanding of population limits.

    - Equation: $\frac{dP}{dt} = rP\left(1 - \frac{P}{K}\right)$, where P is the population, r is the growth rate, and K is the carrying capacity.

3. Logistics

- **Traveling Salesman Problem**:

    - A salesman must visit multiple cities while minimizing travel distance.

    - **Solution**:

        - Use optimization algorithms to find the shortest possible route.

- **Supply Chain Management**:

    - Optimize inventory levels to balance costs and demand.

    - **Example**:

        - The Economic Order Quantity (EOQ) model minimizes total inventory costs: $EOQ = \sqrt{\frac{2DS}{H}}$, where D is demand, S is ordering cost, and H is holding cost.

. . . .

INTRODUCING CALCULUS TO ENHANCE PROBLEM-SOLVING

1. Role of Calculus in Optimization

Calculus provides tools to find maximum and minimum values of functions by analyzing their rates of change.

- **Critical Points**:

  - A critical point occurs where the derivative of a function is zero or undefined: $f'(x) = 0$ or $f'(x)$ undefined.

  - These points are candidates for maxima or minima.

- **Second Derivative Test**:

  - Determine the nature of a critical point: $f''(x) > 0$ local minimum or $f''(x) < 0$ local maximum.

2. Lagrange Multipliers for Constrained Optimization

- **Method**:

  - To optimize $f(x,y)$ subject to a constraint $g(x,y) = 0$, $\nabla f = \lambda \nabla g$, where λ is a scalar called the Lagrange multiplier.

- **Example**:

  – Maximize the area of a rectangle $A = l \cdot w$ given a fixed perimeter.

3. Application of Calculus in Real Problems

- **Example**:

  – Maximize the volume of a box with a given surface area.

  – **Step 1**: Develop the volume function $V(x)$ in terms of dimensions.

  – **Step 2**: Use constraints to express one variable in terms of the others.

  – **Step 3**: Differentiate $V(x)$ and find critical points to optimize.

• •

VISUALIZATION AND COMPUTATIONAL TOOLS

1. Graphical Representation

- Visualizing functions helps identify trends and approximate maxima and minima.

- Tools like graphing calculators and software (e.g., MATLAB, GeoGebra) simplify this process.

2. Computational Optimization

- Algorithms and programming languages like Python provide efficient solutions for complex problems.

- Libraries such as SciPy and NumPy include built-in functions for optimization tasks.

• •

CONCLUSION

Optimization is a cornerstone of problem-solving, enabling us to make the best decisions in various contexts. From minimizing travel distances to maximizing profits or areas, optimization blends geometry, algebra, and calculus into a unified framework for real-world applications. By exploring the mathematical principles and tools behind optimization, we uncover its profound impact across disciplines like economics, biology, and logistics, showcasing its importance in shaping a more efficient and effective world.

CHAPTER 13:
COMPLEX NUMBERS
AND GEOMETRY

• •

Complex numbers bridge the realms of algebra and geometry, offering a robust framework to represent and solve problems involving two-dimensional systems. From their algebraic properties to their geometric representation on the Argand plane, complex numbers provide powerful tools for analysis and application in fields like electrical engineering and quantum physics. This chapter explores the fundamentals of complex numbers, their geometric interpretations, and their practical applications.

• •

COMPLEX NUMBERS AND THEIR PROPERTIES

1. Definition of Complex Numbers

- **Form**:

    - A complex number is written as: $z = a + bi$

        - a: Real part ($\text{Re}(z) = a$).

        - b: Imaginary part ($\text{Im}(z) = b$).

        - i: Imaginary unit, defined as $i^2 = -1$.

2. Fundamental Properties

- **Addition**:

    - Add real parts and imaginary parts separately: $(a + bi) + (c + di) = (a + c) + (b + d)i$

- **Multiplication**:

    - Expand using the distributive property: $(a + bi)(c + di) = (ac - bd) + (ad + bc)i$

- **Conjugate**:

    - The conjugate of $z = a + bi$ is $\bar{z} = a - bi$.

- Used to simplify division: $\dfrac{z_1}{z_2} = \dfrac{z_1 \cdot \overline{z_2}}{|z_2|^2}$

- **Modulus:**

  - The modulus of $z = a + bi$ is: $|z| = \sqrt{a^2 + b^2}$

  - Represents the distance of z from the origin on the Argand plane.

3. Polar Form of Complex Numbers

- **Conversion:**

  - A complex number can also be expressed in polar form: $z = r(\cos\theta + i\sin\theta)$

    - $r = |z|$: Modulus (distance from origin).

    - $\theta = \arg(z)$: Argument (angle with the positive real axis).

- **Exponential Form:**

  - Using Euler's formula $e^{i\theta} = \cos\theta + i\sin\theta$, the polar form becomes: $z = re^{i\theta}$

· ·

GEOMETRIC REPRESENTATION ON THE ARGAND PLANE

1. The Argand Plane

- **Definition**:

  - A Cartesian plane where:

    - The x-axis represents the real part ($Re(z)$).

    - The y-axis represents the imaginary part ($Im(z)$).

- **Plotting**:

  - A complex number $z = a + bi$ is represented as the point (a,b) or as a vector from the origin to (a,b).

2. Geometric Operations

- **Addition and Subtraction**:

  - Vector addition: Add corresponding coordinates.

  - Subtraction: Perform vector subtraction.

- **Multiplication**:

  - Scaling and rotation:

    - Multiplying by $re^{i\theta}$ scales the modulus by r and rotates the number by θ.

- **Division**:

  - Dividing by $re^{i\theta}$ scales the modulus by $\frac{1}{r}$ and rotates the number by $-\theta$.

3. Applications of the Argand Plane

- **Roots of Unity**:

  - The n-th roots of unity are evenly spaced points on the unit circle.

  - For $z^n = 1$, the roots are: $z_k = e^{i\frac{2\pi k}{n}}$, $\quad k = 0,1,\ldots,n-1$

- **Loci of Complex Numbers**:

  - Geometric constraints on complex numbers define curves and regions on the Argand plane, such as circles or lines.

• •

APPLICATIONS IN ELECTRICAL ENGINEERING

1. Alternating Current (AC) Analysis

- **Voltage and Current**:

    - AC signals are represented as complex phasors: $V(t) = V_0 e^{i\omega t}, \quad I(t) = I_0 e^{i(\omega t + \phi)}$

        - V_0: Voltage amplitude.

        - ω: Angular frequency.

        - ϕ: Phase difference.

- **Impedance**:

    - Impedance Z is a complex quantity combining resistance R and reactance X: $Z = R + iX$

        - Used to analyze circuits with resistors, capacitors, and inductors.

2. Power in AC Circuits

- **Power Factor**:

    - The real power P is related to the apparent power S by the cosine of the phase angle ϕ: $P = S\cos\phi$

- **Phasor Diagrams**:

  - Visualize the relationship between voltage and current using the Argand plane.

3. Resonance

- **Application**:

  - Resonant frequencies in circuits are calculated using complex impedance and its dependence on frequency.

• •

APPLICATIONS IN QUANTUM PHYSICS

1. Wavefunctions

- **Complex Representation**:

  - Quantum wavefunctions are often complex: $\psi(x,t) = Ae^{i(kx - \omega t)}$

    - A: Amplitude.

    - k: Wave number.

    - ω: Angular frequency.

- **Probability**:

  - The square of the modulus of Ψ gives the probability density: $|\psi|^2 = \psi \cdot \overline{\psi}$

2. Schrödinger Equation

- **Complex Numbers in Dynamics**:

  - The time-dependent Schrödinger equation uses complex functions to describe quantum states: $i\hbar \frac{\partial \psi}{\partial t} = H\psi$

    - Solutions involve eigenvalues and eigenfunctions represented in complex form.

3. Quantum Interference

- **Wave Superposition**:

  - Complex numbers describe the constructive and destructive interference of quantum waves, central to understanding phenomena like the double-slit experiment.

· ·

ADVANCED
APPLICATIONS

1. Signal Processing

- Complex numbers are used in Fourier transforms, essential for analyzing signals in both time and frequency domains.

2. Control Systems

- Stability of systems is analyzed using the poles and zeros of complex transfer functions on the Argand plane.

3. Fluid Dynamics

- Complex potential functions model flow patterns around objects, solving problems in aerodynamics and hydrodynamics.

• •

CONCLUSION

Complex numbers unify algebra and geometry, offering a versatile framework for solving multidimensional problems. Their representation on the Argand plane bridges theoretical and practical mathematics, making them indispensable in fields like electrical engineering and quantum physics. As tools for analyzing oscillatory phenomena, modeling wave behavior, and optimizing systems, complex numbers continue to illuminate the hidden intricacies of science and mathematics.

CHAPTER 14: MATHEMATICAL MODELING ACROSS DISCIPLINES

• •

Mathematical modeling translates real-world problems into mathematical frameworks, allowing for analysis, prediction, and optimization. By integrating algebra, geometry, and trigonometry, models become powerful tools to address challenges across various fields. This chapter explores how these branches of mathematics work together, illustrates applications through case studies, and highlights interdisciplinary projects that combine their strengths.

• •

THE FOUNDATION OF MATHEMATICAL MODELING

1. What is Mathematical Modeling?

- **Definition**: Mathematical modeling is the process of representing real-world phenomena with mathematical equations and relationships.

- **Steps in Modeling**:

  1. **Problem Identification**:

     - Define the system and identify key variables.

  2. **Formulation**:

     - Develop equations or geometric relationships to represent the problem.

  3. **Analysis and Solution**:

     - Solve equations or analyze the model.

  4. **Interpretation and Validation**:

     - Compare results with real-world data and refine the model.

  5. **Application**:

     - Use the model to predict or optimize outcomes.

2. Role of Algebra, Geometry, and Trigonometry

- **Algebra**:

- Solves equations and represents relationships between variables.

- **Geometry**:
  - Visualizes shapes, spaces, and spatial relationships.

- **Trigonometry**:
  - Models oscillatory and angular phenomena, such as waves and rotations.

• •

CASE STUDIES IN MATHEMATICAL MODELING

1. Bridge Design

- **Problem**: How to ensure a bridge is structurally sound, stable, and cost-efficient.

- **Mathematical Concepts**:

    - **Geometry**:

        - Triangles in trusses distribute forces evenly.

        - Parabolic curves are used in suspension bridges.

    - **Algebra**:

        - Systems of equations calculate load distribution.

    - **Trigonometry**:

        - Models the angle of forces on cables and beams.

- **Example**:

    - **Golden Gate Bridge**:

        - Suspension cables follow a parabolic shape modeled by: $y = ax^2 + bx + c$

        - Engineers calculate tensions using trigonometric ratios and the Pythagorean theorem.

• •

2. Planetary Motion

- **Problem**: Understanding and predicting the orbits of celestial bodies.

- **Mathematical Concepts**:

  - **Algebra**:

    - Kepler's third law relates orbital period (T) and semi-major axis (a): $T^2 \propto a^3$

  - **Geometry**:

    - Orbits are elliptical, described by: $\dfrac{x^2}{a^2} + \dfrac{y^2}{b^2} = 1$

  - **Trigonometry**:

    - Models the angular velocity and position over time. $\theta(t) = \omega t$

- **Example**:

  - Modeling the orbit of Mars helped Johannes Kepler refine the laws of planetary motion.

• •

3. Data Visualization

- **Problem**: Representing complex datasets visually for better analysis and decision-making.

- **Mathematical Concepts**:

  - **Algebra**:

    - Regression models fit lines or curves to data points. $y = mx + b$ linear regression

  - **Geometry**:

- Graphical plots use Cartesian coordinates to display relationships.
  - **Trigonometry**:
    - Polar plots represent cyclic phenomena, such as seasonal variations.
- **Example**:
  - **COVID-19 Spread Visualization**:
    - Logarithmic and exponential functions modeled infection rates.
    - Graphs and heatmaps showed trends and clusters.

• •

INTERDISCIPLINARY PROJECTS COMBINING ALL THREE BRANCHES

1. Climate Modeling

- **Challenge**: Predict climate patterns and their effects on ecosystems.

- **Mathematical Integration**:

  - **Algebra**:

    - Solves differential equations for temperature, precipitation, and wind patterns.

  - **Geometry**:

    - Models Earth's surface and atmospheric layers as spherical shells.

  - **Trigonometry**:

    - Tracks solar angles to model diurnal and seasonal temperature variations.

- **Application**:

  - Predicting the impact of greenhouse gas emissions on global temperatures.

• •

2. *Urban Planning*

- **Challenge**: Optimize the layout of roads, utilities, and buildings.

- **Mathematical Integration**:

  - **Algebra**:
    - Models traffic flow and utility distribution.

  - **Geometry**:
    - Designs spaces and visualizes layouts.

  - **Trigonometry**:
    - Calculates angles and distances in construction projects.

- **Application**:

  - Designing transportation systems to minimize congestion and travel time.

. . . .

3. *Robotics and Automation*

- **Challenge**: Program robots for precision tasks, such as assembly or navigation.

- **Mathematical Integration**:

  - **Algebra**:
    - Controls robotic motion using matrix equations and algorithms.

  - **Geometry**:
    - Describes the robot's workspace and movement paths.

  - **Trigonometry**:
    - Calculates joint angles and rotational movements.

- **Application**:

–   Programming robotic arms in automotive
manufacturing.

. . . .

MATHEMATICAL TOOLS FOR MODELING

1. Graphing and Visualization

- Tools like MATLAB, GeoGebra, and Python libraries (e.g., Matplotlib) help visualize complex models.

- Example:

  – Plotting 3D surfaces to represent terrain or stress in materials.

2. Computational Methods

- Optimization algorithms, such as gradient descent, solve large-scale models.

- Numerical methods, like Euler's method, approximate solutions to differential equations.

3. Integration with Machine Learning

- Mathematical models form the basis of algorithms used in AI for prediction and optimization.

- Example:

  – Neural networks use matrix algebra and trigonometric activation functions.

• •

CONCLUSION

Mathematical modeling demonstrates the power of combining algebra, geometry, and trigonometry to address real-world problems. From designing bridges and predicting planetary motion to visualizing data and planning cities, these disciplines work together to provide insights and solutions. As interdisciplinary projects increasingly integrate mathematical tools, modeling becomes essential for innovation and decision-making across fields like engineering, science, and technology.

CHAPTER 15: HISTORICAL EVOLUTION OF TRIGONOMETRIC TABLES

• •

Trigonometric tables are one of the earliest tools developed to facilitate complex calculations involving angles and distances. From ancient civilizations to modern computational methods, the development of trigonometric tables reflects the ingenuity and mathematical advances of humanity. This chapter explores how early cultures calculated trigonometric ratios, the evolution of trigonometric tables through history, and the modern algorithms that make computations faster and more accurate.

• •

HOW ANCIENT CIVILIZATIONS CALCULATED TRIGONOMETRIC RATIOS

1. Early Uses of Trigonometry

- **Egyptians and Babylonians**:

    – Used basic geometric principles for astronomy, surveying, and construction.

    – The Egyptians applied ratios in building pyramids, with implicit use of what we now call tangent and sine.

    – The Babylonians worked with a base-60 system to estimate angles and distances, laying the groundwork for trigonometry.

2. Greek Contributions

- **Hipparchus (190–120 BCE)**:

    – Known as the "Father of Trigonometry."

    – Compiled the first known trigonometric table to calculate the lengths of chords in a circle for different angles.

    – Used a circle with a radius of 60 units, dividing it into

360 degrees.

- Introduced the concept of a chord, which led to modern
  sine functions:
  $$\text{Chord}(\theta) = 2r\sin\left(\frac{\theta}{2}\right)$$

- **Ptolemy (100–170 CE):**

  - Expanded Hipparchus's work in his seminal text, *Almagest*.

  - Developed a table of chords for angles between 0° and 180° in increments of 0.5°.

  - Used interpolation methods to improve accuracy.

3. Indian Contributions

- **Aryabhata (476–550 CE):**

  - Introduced the sine function (called *jya* or *ardhajya*, meaning "half-chord").

  - His trigonometric tables calculated sine values for angles in increments of 3.75°.

- **Brahmagupta (598–668 CE):**

  - Enhanced sine and cosine tables and introduced the use of tangents.

  - First to define relationships like: $\sin^2(x) + \cos^2(x) = 1$

- **Madhava of Sangamagrama (14th Century):**

  - Part of the Kerala School of Astronomy and Mathematics.

  - Derived series expansions for sine and cosine functions, precursors to modern Taylor series:

$$\sin(x) = x - \frac{x^3}{3!} + \frac{x^5}{5!} - \ldots \quad \cos(x) = 1 - \frac{x^2}{2!} + \frac{x^4}{4!} - \ldots$$

• •

DEVELOPMENT OF TRIGONOMETRIC TABLES THROUGH HISTORY

1. Islamic Golden Age

- **Al-Battani (858–929 CE):**

    – Refined earlier Greek and Indian works.

    – Created sine and tangent tables for astronomy and navigation.

    – Introduced the concept of cotangent.

- **Omar Khayyam (1048–1131 CE):**

    – Calculated trigonometric values with high precision, especially for astronomical calculations.

- **Nasir al-Din al-Tusi (1201–1274 CE):**

    – Compiled highly accurate sine and tangent tables.

    – Used trigonometry to solve spherical geometry problems in astronomy.

2. European Renaissance

- **Regiomontanus (1436–1476 CE):**

    – Published *De Triangulis Omnimodis*, which standardized trigonometric concepts.

- Used trigonometric tables to solve triangles in navigation and astronomy.

- **Bartholomaeus Pitiscus (1561–1613 CE):**

  - Coined the term "trigonometry."

  - Published highly accurate trigonometric tables, including sine, cosine, and tangent values to 10 decimal places.

3. Modern Trigonometric Tables

- **John Napier (1550–1617):**

  - Invented logarithms, simplifying the use of trigonometric tables.

  - Published tables that combined logarithms and trigonometric values, greatly aiding calculations in navigation and astronomy.

- **Edmund Gunter (1581–1626):**

  - Created Gunter's scale, a precursor to the slide rule.

  - His logarithmic trigonometric tables simplified calculations for surveyors and navigators.

- **19th Century Advances:**

  - Charles Babbage and George Airy introduced mechanical devices to compute trigonometric tables.

  - These efforts predated the invention of digital computers but laid the groundwork for automated calculations.

• •

MODERN COMPUTATIONAL METHODS AND ALGORITHMS

1. Numerical Approximation

- Trigonometric functions are computed using series expansions or iterative methods.

- **Taylor Series**:

  - Computes sine and cosine functions by summing an infinite series: $\sin(x) = x - \dfrac{x^3}{3!} + \dfrac{x^5}{5!} - \ldots$

- **Cordic Algorithm**:

  - Iterative method for calculating trigonometric functions, widely used in calculators and embedded systems.

2. Computers and Software

- **Digital Era**:

  - Algorithms like the Fast Fourier Transform (FFT) use trigonometric functions to analyze periodic data in signal processing.

- **Programming Libraries**:

  - Software like MATLAB, Python (NumPy), and

R provides built-in functions for precise trigonometric calculations.

- **High-Performance Computing**:

  – Supercomputers calculate trigonometric values with extreme accuracy, supporting fields like quantum mechanics and climate modeling.

3. Trigonometric Tables in Modern Technology

- Despite the availability of calculators and software, trigonometric tables remain valuable for:

  – Educational purposes to understand trigonometry fundamentals.

  – Applications requiring quick lookup tables, such as in robotics and embedded systems.

· ·

APPLICATIONS OF TRIGONOMETRIC TABLES

1. Astronomy

- Trigonometric tables were indispensable for early astronomers to calculate planetary positions, eclipses, and stellar distances.

2. Navigation

- Mariners and explorers relied on trigonometric tables to determine latitude, longitude, and direction using celestial navigation.

3. Surveying and Engineering

- Tables simplified calculations for measuring distances and angles in construction and land surveying.

4. Modern Signal Processing

- Algorithms rooted in trigonometric principles process audio, video, and other signals in real-time applications like telecommunications and medical imaging.

• •

CONCLUSION

The evolution of trigonometric tables is a testament to humanity's quest for precision and efficiency in mathematics. From ancient Greek chords to modern computational algorithms, trigonometric tables have advanced alongside mathematical thought and technological innovation. Their continued relevance in education, engineering, and computational science underscores their enduring legacy as a foundational tool in mathematics.

CHAPTER 16: CONNECTIONS TO CALCULUS AND LINEAR ALGEBRA

The interplay between calculus, linear algebra, and geometry forms the foundation for solving complex problems in mathematics, engineering, and technology. Calculus adds a dynamic dimension by studying change and accumulation, while linear algebra focuses on vector spaces and transformations. This chapter explores the geometric significance of derivatives and integrals, highlights the role of linear algebra in transformations and 3D modeling, and showcases applications in cutting-edge fields like machine learning and robotics.

INTRODUCTION TO CALCULUS CONCEPTS AND THEIR GEOMETRIC SIGNIFICANCE

1. Derivatives: Rates of Change

- **Definition**:

  - The derivative of a function measures the rate at which
    $$f'(x) = \lim_{h \to 0} \frac{f(x + h) - f(x)}{h}$$
    a quantity changes.

- **Geometric Significance**:

  - The derivative represents the slope of the tangent line to the curve $y = f(x)$ at a point x.

  - A positive derivative indicates an increasing function, while a negative derivative indicates a decreasing function.

- **Applications**:

  - **Maxima and Minima**:

    - Critical points ($f'(x) = 0$) help identify peaks and

troughs in graphs.

- **Motion**:

    - Velocity ($v(t) = s'(t)$) and acceleration ($a(t) = v'(t)$) describe motion along a path.

2. Integrals: Accumulation

- **Definition**:

    - The integral represents the accumulation of quantities, such as area under a curve. $\int_a^b f(x)dx$

- **Geometric Significance**:

    - The definite integral calculates the area under the curve $y = f(x)$ from $x = a$ to $x = b$.

- **Applications**:

    - **Area and Volume**:

        - Used to compute the area of irregular shapes and the volume of solids of revolution.

    - **Physics**:

        - Work done by a force is calculated using: $W = \int F(x)dx$

3. Connections to Geometry

- Derivatives and integrals extend geometry to dynamic systems:

    - Tangent lines represent instantaneous rates of change.

    - Curves are analyzed by integrating their equations.

• •

LINEAR ALGEBRA IN TRANSFORMATIONS AND 3D MODELING

1. Vectors and Matrices

- **Vectors**:

  – Represent direction and magnitude in 2D or 3D space.

  – Example: $v = [x,y,z]$ in 3D space.

- **Matrices**:

  – Represent transformations such as scaling, rotation, and translation.

  – Example:

$$S = \begin{bmatrix} k_x & 0 & 0 \\ 0 & k_y & 0 \\ 0 & 0 & k_z \end{bmatrix}$$

  - Scaling matrix:

2. Linear Transformations

- **Definition**:

  – Linear transformations map vectors to new positions using matrices.

  – Example:

$$R = \begin{bmatrix} \cos\theta & -\sin\theta \\ \sin\theta & \cos\theta \end{bmatrix}$$

- Rotation of a vector v in 2D:

$v' = R{\cdot}v$

- **Geometric Significance**:

  - Transformations can scale, rotate, or shear geometric shapes.

  - Example:

    - Shear transformation affects parallelism but preserves area.

3. Applications in 3D Modeling

- **Rendering 3D Objects**:

  - Use matrices to rotate, scale, and translate 3D models.

  - Example:

    - A 3D point $P = [x,y,z]$ can be transformed by a matrix M: $P' = M{\cdot}P$

- **Projection**:

  - Convert 3D objects to 2D views using projection matrices, such as:

    - Orthographic projection.

$$P' = \begin{bmatrix} 1 & 0 & 0 \\ 0 & 1 & 0 \\ 0 & 0 & 1/d \end{bmatrix}{\cdot}P$$

    - Perspective projection: where d is the distance from the viewer.

APPLICATIONS IN MACHINE LEARNING

1. Feature Representation

-     Linear algebra represents data as vectors and matrices, simplifying operations in machine learning.

- **Example**:

    - A dataset with n features for m samples is represented as a matrix:

$$X = \begin{bmatrix} x_{11} & x_{12} & \cdots & x_{1n} \\ x_{21} & x_{22} & \cdots & x_{2n} \\ \vdots & \vdots & \ddots & \vdots \\ x_{m1} & x_{m2} & \cdots & x_{mn} \end{bmatrix}$$

2. Principal Component Analysis (PCA)

-     PCA uses eigenvalues and eigenvectors to reduce dimensionality while retaining the most significant features.

- **Steps**:

    - Compute the covariance matrix C of the data.

    - Find eigenvectors and eigenvalues of C.

    - Transform data using the principal eigenvectors.

3. Gradient Descent

-     Calculus optimizes machine learning models by

minimizing loss functions.

- **Process:**

  - Update model parameters θ iteratively: $\theta = \theta - \eta \nabla L(\theta)$
    where η is the learning rate, and $\nabla L(\theta)$ is the gradient of the loss function.

• •

APPLICATIONS
IN ROBOTICS

1. Kinematics

- **Forward Kinematics**:

    - Use matrices to calculate the position of a robot's end effector based on joint angles.

    - Example:

        - A robotic arm with two joints: $P = T_1 \cdot T_2 \cdot P_0$

        where T_1 and T_2 are transformation matrices for the joints.

- **Inverse Kinematics**:

    - Solve for joint angles given a desired position of the end effector.

    - Requires iterative methods using calculus.

2. Path Planning

- Linear algebra and calculus combine to optimize paths for robots to avoid obstacles and reach targets efficiently.

3. Control Systems

- Represented by differential equations, robot control systems rely on calculus to predict and adjust movements.

· ·

CONCLUSION

The integration of calculus and linear algebra into geometry enriches mathematical modeling and problem-solving across diverse fields. Derivatives and integrals provide dynamic insights into systems, while matrices and transformations enable precise modeling of multidimensional spaces. Applications in machine learning and robotics showcase the practical power of these tools, driving innovation and efficiency in modern technology. The connections between these disciplines highlight the seamless interplay of mathematics in understanding and shaping the world.

CHAPTER 17:
ADVANCED PROBLEM-
SOLVING STRATEGIES

• •

Mathematics challenges us to think critically, explore creatively, and find efficient solutions to complex problems. Advanced problem-solving strategies incorporate heuristic approaches, tackle competitive mathematics problems, and build intuition for interdisciplinary problem-solving. This chapter delves into techniques for breaking down challenging problems, highlights strategies for excelling in competitive mathematics, and fosters skills for tackling real-world, interdisciplinary issues.

• •

HEURISTIC APPROACHES TO TACKLING COMPLEX PROBLEMS

1. What Are Heuristics?

- **Definition**:

  - Heuristics are problem-solving strategies or "rules of thumb" that guide reasoning and decision-making.

  - They provide efficient, though not guaranteed, methods for finding solutions.

2. Common Heuristic Strategies

- **Guess and Check**:

  - Make an educated guess, test its validity, and refine based on feedback.

  - Example:

    - Solve $x^2 + 5x - 24 = 0$ by testing possible integer roots.

- **Divide and Conquer**:

  - Break the problem into smaller, manageable parts.

  - Example:

- Solve a geometry problem by analyzing individual triangles or subregions.

- **Working Backward**:
  - Start with the desired result and reverse the steps to reach the given conditions.
  - Example:
    - In logic puzzles, trace the solution from the conclusion to the initial setup.

- **Pattern Recognition**:
  - Identify patterns or symmetries to simplify problems.
  - Example:
    - Recognize the Fibonacci sequence in a combinatorics problem.

- **Simplify the Problem**:
  - Reduce complexity by solving a simpler version of the problem first.
  - Example:
    - Solve a quadratic equation with small coefficients before tackling larger or symbolic coefficients.

3. Benefits of Heuristics
- Develops intuition for problem structure.
- Encourages iterative thinking and adaptability.
- Provides starting points for otherwise daunting challenges.

• •

SOLVING COMPETITIVE MATHEMATICS PROBLEMS

Competitive mathematics problems are designed to test ingenuity, creativity, and deep understanding of mathematical principles. Excelling requires both theoretical knowledge and strategic thinking.

1. Common Types of Problems

- **Algebra**:

    - Solve equations, inequalities, and functional relationships.

    - Example:

        - Solve for x: $\sqrt{x + 3} + \sqrt{2x + 1} = 4$

- **Geometry**:

    - Analyze shapes, angles, areas, and volumes.

    - Example:

        - Prove that the sum of the angles in a triangle is 180° using parallel lines.

- **Number Theory**:

- Explore properties of integers, divisibility, and modular arithmetic.

- Example:

  - Find the smallest positive integer x such that $7x \equiv 1 \pmod{13}$.

- **Combinatorics**:

  - Solve counting, arrangement, and probability problems.

  - Example:

    - How many ways can 5 people sit in a row such that two specific people are always next to each other?

- **Calculus**:

  - Analyze rates of change and accumulation.

  - Example:

    - Maximize the volume of a box with a fixed surface area.

2. Strategies for Success

- **Understand the Problem**:

  - Carefully read and reframe the problem in simpler terms.

- **Draw Diagrams**:

  - Visual aids help clarify relationships, especially in geometry and graphing.

- **Apply Known Results**:

  - Use theorems, formulas, and previously solved problems as shortcuts.

- **Check Units and Dimensions**:

- Ensure consistency in physical and mathematical quantities.

- **Look for Edge Cases**:

 - Consider extreme or boundary conditions to verify solutions.

- **Proof and Rigor**:

 - Ensure all steps are justified, especially in theoretical proofs.

3. Practice Resources

- Engage with problem sets from Olympiads, contests, and advanced textbooks.

- Use online platforms like Art of Problem Solving, Brilliant.org, or Khan Academy.

- Study past competitions, such as AMC, IMO, or Putnam Exam problems.

• •

BUILDING INTUITION FOR INTERDISCIPLINARY PROBLEM-SOLVING

Mathematics often intersects with other disciplines, requiring a blend of domain knowledge, creativity, and analytical skills.

1. Bridging Mathematics with Other Fields

- **Physics**:
    - Use calculus and differential equations to model motion, forces, and energy.
    - Example:
        - Derive the trajectory of a projectile using $s(t) = v_0 t + \frac{1}{2}at^2$.

- **Biology**:
    - Apply statistics and probability to genetics, ecology, and epidemiology.
    - Example:
        - Model the spread of a virus using exponential growth equations.

- **Economics**:

- Optimize profit or utility functions using derivatives.
- Example:
  - Solve for the price that maximizes revenue given demand: $R(p) = p \cdot q(p)$

- **Computer Science**:
  - Algorithms and data structures rely on combinatorics and linear algebra.
  - Example:
    - Use graph theory to optimize network connections.

2. Developing Intuition

- **Understand Core Principles**:
  - Deepen your understanding of fundamental mathematical tools and their applications.

- **Think Multidimensionally**:
  - Approach problems from different perspectives, using geometry, algebra, or calculus as appropriate.

- **Relate to Real-World Contexts**:
  - Contextualize abstract problems with practical examples.

3. Interdisciplinary Problem-Solving Framework

- **Identify the Problem**:
  - Define variables and constraints clearly.

- **Choose Appropriate Tools**:
  - Determine which mathematical methods apply.

- **Iterate and Refine**:
  - Test solutions, adjust assumptions, and repeat as necessary.

- **Collaborate Across Disciplines**:

  - Engage with experts from relevant fields for deeper insights.

• •

APPLICATIONS
OF ADVANCED
PROBLEM-SOLVING

1. Space Exploration

- Optimize spacecraft trajectories using calculus and differential equations.

- Example:

  – Solve the two-body problem to calculate orbits.

2. Climate Science

- Model weather patterns and predict climate change impacts using numerical methods and linear algebra.

3. Artificial Intelligence

- Train neural networks using gradient descent, which minimizes loss functions iteratively.

4. Medicine

- Apply optimization to design efficient radiation therapy plans.

• •

CONCLUSION

Advanced problem-solving requires a blend of heuristic strategies, rigorous mathematical skills, and interdisciplinary thinking. By tackling competitive problems and applying mathematical intuition to real-world challenges, we expand our ability to analyze and solve complex issues. As mathematics continues to intersect with diverse fields, mastering these strategies prepares us to innovate and excel in an increasingly interconnected world.

CHAPTER 18:
THE FUTURE OF
MATHEMATICS IN
TECHNOLOGY

Mathematics continues to be a driving force behind technological advancement, enabling breakthroughs in artificial intelligence (AI), virtual reality (VR), quantum computing, and sustainable technologies. The foundational branches of algebra, geometry, and trigonometry provide the frameworks for innovation and discovery in these fields. This chapter explores the current and emerging roles of mathematics in cutting-edge technologies, highlights its contributions to sustainability, and speculates on the future of mathematical research.

ROLE OF ALGEBRA, GEOMETRY, AND TRIGONOMETRY IN EMERGING TECHNOLOGIES

1. Artificial Intelligence (AI)

AI leverages mathematical models to replicate human-like intelligence in machines.

- **Algebra in AI**:

  - **Linear Algebra**:

    - Forms the backbone of machine learning and neural networks.

    - **Example**:

      - Representing data as matrices or tensors for computation.

    - **Equation**:

      - A neural network layer can be represented as: $y = \sigma(Wx + b)$, where W is a weight matrix, x is an input vector, b is a bias vector, and σ is the activation function.

- **Optimization**:
  - Calculus-based gradient descent algorithms minimize loss functions to train models.

- **Geometry in AI**:
  - **High-Dimensional Spaces**:
    - Data points are treated as vectors in high-dimensional geometric spaces.
    - Distance metrics like Euclidean distance are used for clustering and classification.
  - **Computer Vision**:
    - Geometry helps in object detection, image reconstruction, and spatial reasoning.

- **Trigonometry in AI**:
  - **Signal Processing**:
    - Fourier transforms, rooted in trigonometry, are used in speech and image recognition.
  - **Recurrent Neural Networks**:
    - Periodic functions like sine and cosine model time-series data.

• •

2. Virtual Reality (VR)

VR systems rely heavily on geometry and trigonometry to create immersive, realistic environments.

- **Geometry in VR**:
  - **3D Modeling**:
    - Objects in VR are created using geometric primitives like cubes, spheres, and polygons.
  - **Rendering**:

- Transformation matrices handle scaling, rotation, and translation of objects.

- **Projection**:

  - Convert 3D coordinates into 2D screen coordinates using perspective projection:
  $$P' = \frac{P}{z},$$ where z is the depth coordinate.

- **Trigonometry in VR**:

  - **Field of View (FoV)**:

    - Calculate viewing angles and distances for realistic rendering.

  - **Lighting and Shading**:

    - Use trigonometric functions to simulate how light interacts with surfaces.

- **Applications**:

  - Gaming, architectural visualization, and medical training simulations.

• •

3. Quantum Computing

Quantum computing exploits the principles of quantum mechanics to perform computations.

- **Algebra in Quantum Computing**:

  - **Linear Algebra**:

    - Quantum states are represented as vectors in complex Hilbert spaces.

    - **Example**:

      - A qubit's state is expressed as:
      $$|\psi\rangle = a|0\rangle + b|1\rangle,$$ where a and b are complex

numbers satisfying $|a|^2 + |b|^2 = 1$.

- **Matrix Operations**:
    - Quantum gates are represented as unitary matrices, which manipulate qubit states.

- **Geometry in Quantum Computing**:
  - **Bloch Sphere**:
    - Visualizes qubit states as points on a sphere, providing an intuitive geometric representation.

- **Trigonometry in Quantum Computing**:
  - **Phase Angles**:
    - Sine and cosine functions represent quantum state rotations.

- **Applications**:
  - Cryptography, optimization problems, and simulation of quantum systems.

• •

MATHEMATICS IN SUSTAINABLE TECHNOLOGIES

1. Renewable Energy Systems

- **Solar Panels**:

  - Trigonometry calculates the optimal angle for solar panel installation to maximize energy absorption.

- **Wind Turbines**:

  - Algebra and geometry model turbine blade efficiency and energy output.

2. Smart Grids

- **Linear Algebra**:

  - Optimizes power distribution across networks.

- **Differential Equations**:

  - Models dynamic energy flow and grid stability.

3. Climate Modeling

- **Geometry**:

  - Models the Earth's surface and atmospheric layers for climate simulations.

- **Numerical Methods**:

  - Solves equations governing weather patterns and global warming trends.

4. Sustainable Architecture

- **Geometry and Optimization**:

  - Optimize building shapes for energy efficiency.

- **Simulation**:

  - Trigonometry and calculus analyze natural light and airflow in buildings.

• •

SPECULATIONS ON THE FUTURE OF MATHEMATICAL RESEARCH

1. Bridging Discrete and Continuous Mathematics

- Development of hybrid approaches to tackle problems involving both discrete and continuous systems, such as modeling biological systems or urban dynamics.

2. Rise of Computational Mathematics

- **Automation of Proofs**:

    - Algorithms to verify and discover mathematical proofs.

- **AI in Mathematics**:

    - AI systems capable of conjecturing and proving theorems.

3. Expansion of Interdisciplinary Research

- **Mathematics and Biology**:

    - Advancements in modeling genetic networks and cellular processes.

- **Mathematics and Neuroscience**:

    - Unraveling the brain's complexity using differential equations and network theory.

4. Quantum Mathematics

- Development of entirely new mathematical frameworks to support quantum computation and quantum communication.

5. Exploration of Higher Dimensions

- Advancements in geometry and topology to understand high-dimensional spaces with applications in string theory and cosmology.

• •

CONCLUSION

Mathematics is the cornerstone of technological innovation, with algebra, geometry, and trigonometry playing pivotal roles in AI, VR, quantum computing, and sustainable technologies. As technology evolves, so too will mathematics, adapting to new challenges and inspiring breakthroughs across disciplines. The future of mathematical research promises to deepen our understanding of the universe while driving progress in engineering, science, and society.

CONCLUSION

Mathematics is a timeless language that transcends boundaries, offering profound insights into the natural world and enabling solutions to complex problems. Throughout this journey, we have explored the interconnected realms of algebra, geometry, and trigonometry, discovering how these disciplines weave together to form the foundation of modern science, technology, and innovation.

THE INTERCONNECTED NATURE OF ALGEBRA, GEOMETRY, AND TRIGONOMETRY

Algebra, geometry, and trigonometry are not isolated fields; they are deeply intertwined, each enhancing the other's scope and capabilities. Algebra provides the symbolic framework for expressing relationships and solving equations, geometry visualizes these relationships in space, and trigonometry quantifies angles and periodic phenomena. Together, they create a robust mathematical toolkit capable of addressing challenges across diverse domains.

- **Examples of Interconnection**:

  - **Coordinate Geometry**: Combines algebra and geometry to describe shapes and trajectories in a Cartesian plane.

  - **Trigonometric Identities in Geometry**: Illuminate properties of circles, triangles, and waves.

  - **Algebraic Equations in Physics**: Relate geometric principles to the behavior of objects in motion, with trigonometry capturing oscillatory systems like sound and light.

By understanding their synergy, we unlock powerful ways to

model, analyze, and innovate in fields ranging from architecture to quantum computing.

• •

ENCOURAGING READERS TO CONTINUE THEIR MATHEMATICAL JOURNEY

Mathematics is a lifelong journey of exploration and discovery. While this book provides a foundation, it is only the beginning. As you advance, consider the following ways to deepen your mathematical understanding:

1. **Engage with Real-World Applications**:

   - Explore how mathematics solves tangible problems in engineering, medicine, economics, and technology.

   - Example: Investigate how trigonometry aids in satellite navigation or how geometry shapes architectural masterpieces.

2. **Participate in Collaborative Projects**:

   - Join interdisciplinary teams that use mathematics to tackle global issues like climate change, healthcare access, and sustainable energy solutions.

3. **Explore Advanced Fields**:

   - Dive into advanced topics like calculus, linear algebra, differential equations, or number theory to broaden your mathematical toolkit.

4. **Stay Curious**:

–   Approach mathematical problems with curiosity and creativity.

–   Question established theories and seek new ways to apply mathematical principles.

By nurturing a passion for mathematics, you will find yourself better equipped to navigate the complexities of an ever-changing world.

• •

MATHEMATICS: A TOOL TO UNDERSTAND THE UNIVERSE

Mathematics is more than an academic discipline; it is a lens through which we perceive the universe. It explains the motions of celestial bodies, the behavior of subatomic particles, and the patterns of ecosystems. Mathematics reveals the hidden symmetry in nature, the elegance of fractals, and the profound order underlying chaos.

- **Examples of Universal Insights**:

  - **Einstein's Relativity**: Relies on differential geometry to describe spacetime.

  - **Quantum Mechanics**: Uses complex numbers and linear algebra to explain wave-particle duality.

  - **Cosmology**: Applies trigonometry and calculus to model the expansion of the universe.

Mathematics also reflects the creativity of the human mind, inspiring art, music, and architecture. From the Fibonacci sequence in sunflower spirals to the parabolic curves in Gothic cathedrals, mathematics connects us to both the physical and the metaphysical.

• •

SOLVING GLOBAL CHALLENGES WITH MATHEMATICS

The world faces unprecedented challenges—climate change, energy shortages, pandemics, and technological ethics. Mathematics provides the tools to address these crises by modeling systems, optimizing resources, and innovating solutions.

- **Applications in Action**:

  - **Epidemiology**: Models disease spread to inform public health policies.

  - **Renewable Energy**: Optimizes the efficiency of solar panels and wind turbines.

  - **Artificial Intelligence**: Shapes ethical algorithms for autonomous systems.

Mathematics empowers us not only to understand these challenges but also to act decisively and effectively.

• •

A FINAL REFLECTION

Mathematics is a unifying force, connecting the abstract with the tangible, the theoretical with the practical, and the logical with the creative. It equips us with the tools to explore, innovate, and solve. As you continue your mathematical journey, remember that each equation, theorem, or concept you master contributes to a greater understanding of the world and your ability to shape its future.

The path of mathematics is infinite—full of beauty, discovery, and purpose. Whether solving equations or unraveling the mysteries of the universe, your journey is part of a legacy of human ingenuity that spans millennia and will continue for generations to come.

www.ingramcontent.com/pod-product-compliance
Lightning Source LLC
Chambersburg PA
CBHW071455220526
45472CB00003B/803